Management of Southwestern Desert Soils

Management of

Southwestern Desert Soils

WALLACE H. FULLER

Author of *Soils of the Desert Southwest*

THE UNIVERSITY OF ARIZONA PRESS
Tucson, Arizona

About the Author . . .

WALLACE H. FULLER has authored numerous scientific articles in national and international journals and is co-author of, or contributor to, several books. He frequently has served as consultant and advisor in biochemistry and in the management of water and soils. A Fellow in the American Association for the Advancement of Science, the Soil Science Society of America, and the American Society of Agronomy, he won the E. B. Knight Award of the National Association of Colleges and Teachers in Agriculture for the best-written article in 1972. A biochemist and professor of soil science, he joined the University of Arizona faculty in 1948, later to serve as department head of Agricultural Chemistry and Soils for sixteen years. Previously he was with the Agricultural Research Service and the Soil Conservation Service of the U. S. Department of Agriculture, with Washington State University, and with Iowa State University. Fuller, a native of Old Hamilton, Alaska, received his B.S. and M.S. degrees from Washington State University and his Ph.D. from Iowa State University.

THE UNIVERSITY OF ARIZONA PRESS

I.S.B.N.-0-8165-0442-3
L. C. No. 74-15601

To Winifred and Pamela

Contents

Illustrations

Tables

Speak to the earth
and it shall teach thee.
Job 12:8

Acknowledgments

Deep appreciation is extended to those who have so generously helped in the preparation of this volume. At the University of Arizona, facilities and other support provided by the Agricultural Experiment Station were invaluable. Critical reviews by Thomas C. Cooper of the Journalism Department and by Arthur W. Warrick and Kenneth K. Barnes of the Department of Soils, Water, and Engineering, were highly beneficial. The encouragement and advice of Kenneth K. Barnes, head of the latter department, was most heartening. The patience and help afforded by my wife, Winifred, played an especially important part in bringing this book to completion. I also wish to express my thanks to Gale Monson for helpful editing and to other personnel at the University of Arizona Press for their role in preparing *Management of Southwestern Desert Soils* for publication.

W.H.F.

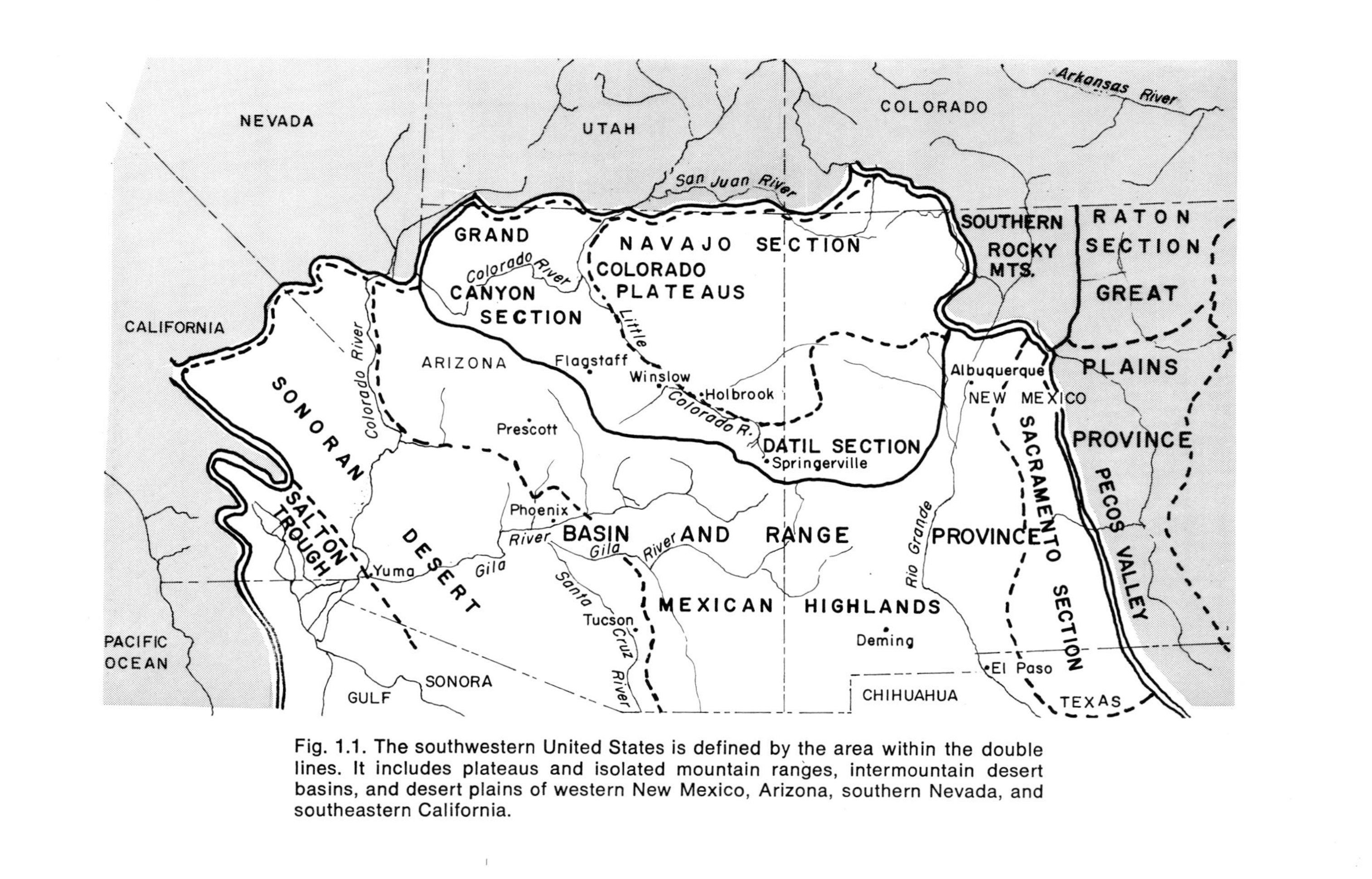

Fig. 1.1. The southwestern United States is defined by the area within the double lines. It includes plateaus and isolated mountain ranges, intermountain desert basins, and desert plains of western New Mexico, Arizona, southern Nevada, and southeastern California.

1. Why Soil Management?

The amount of desert land suitable for plant production has been diminished by man's invasion. Let's take a look: city and town buildings encroach on rich residential gardens; mining either strips or covers fertile soil; highways, airports, and parking lots pave the land with asphalt; and not the least factor in the spoilage of land is its mismanagement. Mismanagement is particularly obvious in desert climates where poor drainage, high water tables, inadequate provision for leaching, bad irrigation water, and lack of understanding of basic principles involved in irrigated plant growing have taken land out of production or made it marginal in quality. Farm cultivation and overgrazing often expose land to accelerated erosion where topsoil and moisture are lost. Dust blown from dry, cultivated land has been so dense at times as to lower highway visibility to zero, causing vehicular accidents and death.

The migration of populations from temperate-humid climates to the sunny-dry deserts of the West has reached unbelievable figures since air conditioning and other comforts have been readily available. Desert living is a new experience for most people. Growing plants in a hot arid climate and in alkaline soils packed with lime and salts is even stranger to the new resident. Indeed, it is strange to some old-timers also. Special practices of soil, water, and plant management are more essential in arid than in humid climates if soils are to continue to support the desired plant growth and quality of living.

Despite the many problems involved in desert plant culture, the studied application of water and fertilizer to soils of the Southwest has raised the food and fiber yields of thousands of acres from scrub vegetation and cactus to some of the highest production levels in the United States, proving that desert land can be made to respond favorably to man's invasion. Many residential areas bloom the year round with exotic plants in a beauty unexceeded by those of other climates. This plant production reflects management practices compatible with permanently productive soils and a continued satisfying way of life. It stands in remarkable contrast to abandoned, mismanaged, salty, and eroded lands.

Land management in the desert Southwest takes into account the *water* and the *plant*, as well as the *soil*. Since soils are not the same, but vary according to their parent material and weathering processes, their support of successful home and farm plantings also differs over a wide range. Correct management is intended to place the soil at the top of its capabilities. Correct management is not aimed at exploiting the immediate productivity of soil, but at a permanency of production. In arid climates, soils are under stresses of irrigation water that is higher in salts than rainwater, of a high evaporation rate from surfaces with concurrent upward movement of salts by capillary action, and of high water demand by plants to accommodate heavy transpiration losses under desert conditions.

Fig. 1.2. An aerial view of typical plant distribution patterns, drainageways, and low mountain cones in the Great Basin area of Arizona and California.

On the other hand, the control of plant growth and production factors is more feasible under irrigation conditions of the desert Southwest than under humid conditions in which more production factors are left to chance.

The interdependence of factors that affect plant growth is not casual. Should any one of the factors not be optimally available, the effectiveness of the other factors is lessened. Good management attempts to keep all the factors as optimum as possible. However, the necessity for optimizing *all* factors is the most difficult concept for growers to understand and put into practice. Those who do, find growing plants in the desert Southwest rewarding.

In the desert Southwest, identified as lying between the double lines in Figure 1.1, most irrigated agricultural areas and sites of major cities are located on the inner-valley flood plains. The soils are deep and, for the most part, well-drained. The general physiography favors soils of high productivity, with gentle slopes that provide a favorable elevation differential for maintaining good drainage of the better soils (Fig. 1.2). Less productive land common to the desert comprises such physiographic types as "alkali flats," undulating to gently rolling plains, sloping alluvial fans and basins, dissected rolling plateaus, moderate to steep hills and mountain footslopes, and steep mountains.

2. How Plants Grow in a Desert

At a southwestern flower show, one particular man took a fifth straight win in the best display of chrysanthemums. Yet he claimed to have no special plan of soil or water management. But in describing that he had selected a deep soil, spaded in well-rotted compost to a depth of 12 inches, soaked the soil to get deep water penetration, used new healthy container-plants, added a little top dressing of ammonium phosphate, never allowed the soil to dry out completely, and mulched the surface with bark compost . . . he narrated an excellent plan of soil management. The prizes testified as much for good soil management as for beautiful plants. The many inquiries of fellow exhibitors were aimed at his secrets of the way soil, water, and fertilizer were managed. Exhibitors knew that his chrysanthemums had responded to some special treatment, and their perfection in growth reflected the success of that treatment.

Requirements for Growth

Plants have certain requirements for growth. The primary aim of soil management in the desert is the improvement of that growth. From the atmosphere plants need *light* (except for some primitive plants), *air* (oxygen and carbon dioxide), and *water*. From the soil the roots need *air* (oxygen and carbon dioxide), *water* (rain or irrigation), and *essential mineral elements* (sometimes called nutrients). The soil also should provide *anchorage*. Favorable air and soil *temperature* is especially emphasized in the desert where temperatures can be excessively hot in the summer.

Plant ecologists of the desert, furthermore, point out the necessity of having the root feeding zone free of *toxic elements* such as boron, and of excessive *salts*, if successful plantings are to be sustained.

These growth requirements or factors are interdependent, and they may be satisfied in many different ways. Best growth occurs when all the requirements become optimum. A deficiency of any one of them can seriously inhibit growth, even though others are optimal. This is sometimes called "the principle of limiting factors." Deficiencies of two or more of the growth factors only compound the adverse growth effect. More often than not, multiple deficiencies result either in the plant not growing at all, or in its prompt death upon germination.

Fortunately, plants adapt to limited deficiencies and excesses naturally. In the desert Southwest where water is the most limiting single factor, vegetation can be divided into three types,* based on water requirements: (1) vegetation dependent upon local precipitation only, (2) vegetation dependent upon accumulation in drainage courses and depressions of rain that falls within the local areas, and (3) vegetation dependent upon moisture from sources outside the desert itself, such as rivers or lakes fed partly or mainly from without the area.

Thus plants show great capacities to adapt to a wide range of soil moisture conditions. Plants classed as 1 and 2 have developed drought-escaping, drought-evading, drought-enduring, and drought-resisting mechanisms that permit them to survive and live in a desert climate. Plants without these mechanisms or adaptations do not invade the desert and die when brought into such a climate, unless irrigated with water from deep underground sources or from without the area.

Not only is desert vegetation adapted to low moisture levels in soils for prolonged periods, but some plants are adapted to high salt concentrations as well. Agriculturally economic crops and residential ornamentals also vary in tolerance to salt. The ecological adaptation of plants is such that some will not grow at all at a given salt level in the soil, whereas others will grow very well, even optimally. The same holds true for tolerance to specific elements (ions) which may accumulate in the soil, like boron, lithium, and bicarbonates.

*McGinnies, William G. 1967. An inventory of geographical research on desert environments: IV. Inventory of research on vegetation of desert environments. Univ. of Arizona Press, Tucson.

Environmental adaptations to soil deficiencies (or very slow availability) of certain plant-required elements also occur with desert vegetation and help to form ecological plant population patterns. For example, the presence of a high amount of lime, or *caliche* (cuh-lee-chee), in soil results in a lime-induced chlorosis in some plants. The deficiency is thought to be a lack of available soil iron, since soluble iron compounds correct the yellowing symptoms noticeable in plant leaves. Variable adaptation occurs within certain species of such agricultural plants as grapes and grain sorghum, and in fruit trees like peaches and plums, as well as in ornamentals and lawn grasses. Broadleaf evergreens (camellia, gardenia) and berry plants (strawberry and blackberry), not native to the desert, find one or more of the micronutrients (iron, zinc, manganese) unavailable, or poorly available, to them.

Plant special adaptations. Proper soil management must take into account the rooting habits of plants and the depth they extend into soil. Plants differ considerably as to the volume of soil they penetrate. Root distribution determines the *moisture-extraction pattern*. The kind of root development — for example, shallow, deep tap, fibrous — is fixed by heredity. Lettuce extends roots only a foot or two, whereas tomatoes feed as deep as three feet. Pansies have a shallow, fibrous root system of 12 to 18 inches, and larkspur will use moisture from 36 inches. Shrubs and trees root deeper, usually four to eight feet if the soil is permeable.

Creosote bush, for example, will send roots to a depth of eight to ten feet where the soil is suitably loose and penetrable. The large volume of soil provides small amounts of water through long periods of drought and keeps the plant alive by the moisture in the soil pore spaces. The humidity of air in the soil approaches 100 percent saturation. Thus the environment is such that the life of the root can be sustained for long periods of dormancy, *providing the demand of the tops is restricted*. Ocotillo, as well as creosote bush, cactus, and yucca, for example, store moisture in their parts sufficient to bloom, even though rain has not fallen for months. Some of the necessary moisture is derived by absorption from the water vapor of the soil pore space.

Plants vary in *rate of transpiration*. By transpiration we mean the removal of water from a soil by roots, and its passage up through the stem, into the leaves, and out the stomata into the atmosphere. *Evaporation* is the loss of water to the atmosphere from the soil surface. Native desert plants use several means of conserving moisture by reducing transpiration losses. For example, water demand for transpiration is lessened by: reduced leaf surface area; thin scales of leaves or no true leaves as we think of them; protective coating on leaves such as wax, resins, and chitin; pubescence, fuzz, or hair on leaves; reversion to dormancy; short life cycle adapted to rainy seasons or even a single shower, and, not the least; an extensive fibrous root system.

The usual garden plants and vegetables familiar to us in humid climates require considerably more water to survive than desert plants. This makes it doubly important to provide a root feeding volume or reservoir as favorable as possible for such plants in desert areas to extract moisture and mineral nutrients. Plants of humid climates develop a more extensive leaf-surface area in relation to the root than plants of arid climates. The purpose is to catch more sunlight for photosynthesis where sunny days appear less frequently than in the desert. Where these plants are placed under hot-dry conditions, the moisture absorbing volume of the roots is the first line of defense for survival.

Plant growth is affected by numerous factors, only some of which concern *fertility* and *water*. So let's look at *all* those factors relating to the capacity of soils to sustain plant growth. They are grouped for convenience into three categories: *climatic*, *biotic*, and *edaphic*. *Climatic* factors which make plants grow differently in soils of arid than in soils of humid areas are primarily low rainfall and humidity, high temperatures, and the many sunny days. *Biotic* (biological) factors include microorganisms (disease and non-disease organisms that influence a plant's welfare), insects, weeds, many animals including man, and the plant itself. *Edaphic* factors are soil properties broadly lumped under chemical, physical, and biological characteristics which influence a soil's capacity to supply nutrients, water, and oxygen (Table 2.1).

The Climate

Climate is the dominant factor affecting plant growth in the desert Southwest. Landscapings and home gardens require water supplemental to that of the scant rainfall. Established native plants and cacti survive, but planted turf, trees, shrubs, and garden plants not native to the desert do not. Rainfall is meager, and evapotrans-

Table 2.1

Factors in Soil Productivity and Fertility

Factors in soil productivity	Factors in soil fertility (essential plant nutrients)
1. Climatic	1. Macronutrients in soil
rainfall	nitrogen (N)
temperature	phosphorus (P)
sunlight	potassium (K)
	calcium (Ca)
2. Biotic	magnesium (Mg)
microorganisms	sulfur (S)
weeds	carbon (C)
insects	oxygen (O)
other animals	hydrogen (H)
man	
	2. Micronutrients in soil
3. Edaphic	iron (Fe)
soil chemistry	manganese (Mn)
soil microbiology	molybdenum (Mo)
soil physics	boron (B)
soil nutrients	copper (Cu)
water	zinc (Zn)
air (oxygen)	chlorine (Cl)
	sodium (Na)
4. Excesses	
elimination of excesses such as salt, toxic elements, water in water-logging.	

piration is high (Figs. 2.1 and 2.2). The air humidity is extremely low and the heat factor high. Rain in the desert Southwest is about equally divided between midwinter and midsummer. The desert Southwest has two wet seasons and two dry seasons (Fig. 2.3). The split wet season lessens the effectiveness of the rain for completing the growth cycle of many plants and the establishment of permanent landscaping or home gardens. Much of the rain evaporates; only a part percolates into the soil. Water must be supplied and managed to supplement any moisture which rain brings to the arid soils. The summer rains fall as violent thunderstorms, often causing serious erosion of soil and water runoff. The water fills washes, arroyos, and streams, sweeping rock, mud, and debris onto productive valley lands and low areas and often damaging poorly located homes and recreation parks and golf courses. Winter rains are less intense and more widespread.

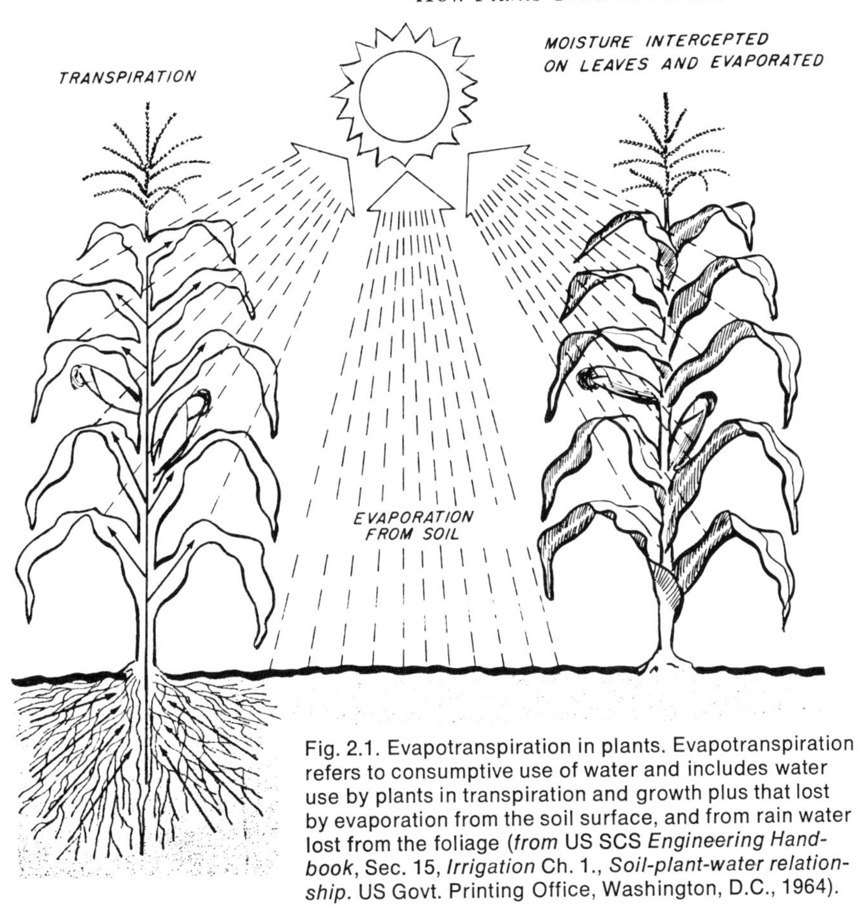

Fig. 2.1. Evapotranspiration in plants. Evapotranspiration refers to consumptive use of water and includes water use by plants in transpiration and growth plus that lost by evaporation from the soil surface, and from rain water lost from the foliage (*from* US SCS *Engineering Handbook*, Sec. 15, *Irrigation* Ch. 1., *Soil-plant-water relationship.* US Govt. Printing Office, Washington, D.C., 1964).

Evaporation of moisture from soil surfaces under a desert climate is high, with consequent accumulation of salts. Soil, water, and cultural management must be imposed to counteract salt accumulation and discourage its capillary rise from lower depths. Since low rainfall does not leach salts as high rainfall does under humid conditions, leaching salts below the root zone must be accomplished by added water, as through irrigation.

The lack of winter refrigeration common to temperate climates allows the soil to lose water over a long period of time as compared to snow-blanketed land, which conserves moisture. Where water is supplied from sources outside of the desert, though, plant growth continues throughout the year. Gardeners become accustomed to

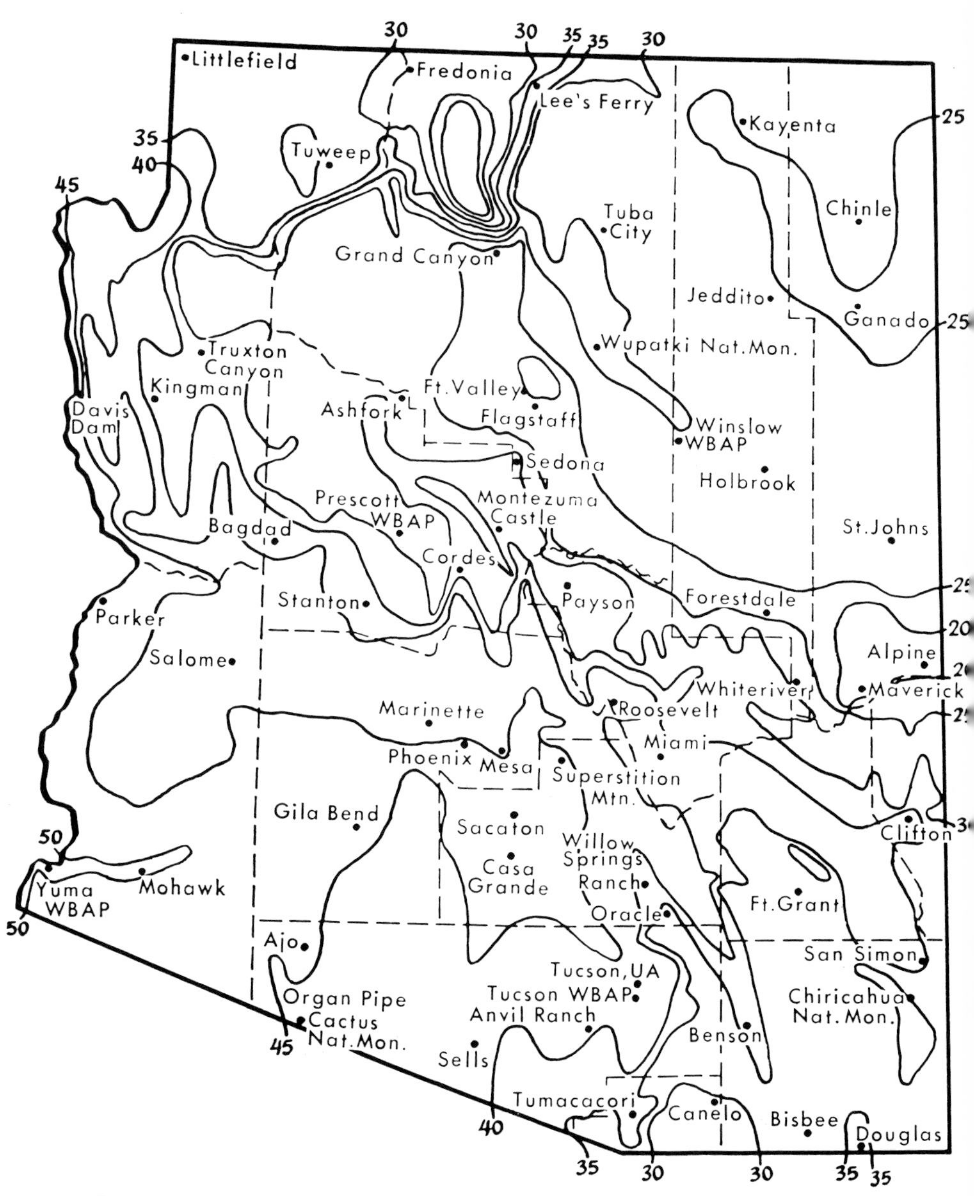

Fig. 2.2. Isograms of annual potential evapotranspiration, in inches, for Arizona (from *Arizona climate*, The Univ. of Arizona Press, Tucson, 1960).

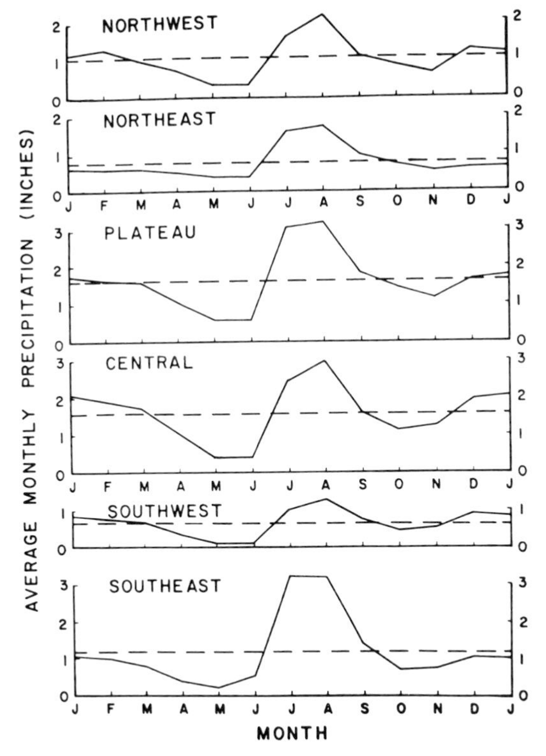

Fig. 2.3. Monthly rainfall distribution pattern in Arizona (from *Arizona climate*, The Univ. of Arizona Press, Tucson, 1960).

planting in the fall. Bedding plants are set out in October and November, and vegetable gardens and flowering annuals are seeded in late September. Winter lawns of annual ryegrass are planted during the first half of October.

As a consequence of low rainfall, high evaporation, and extended periods of drought, the native vegetation is restricted in growth and sparse. The true desert vegetation consists of creosote bush, cactus, yucca, and palo verde, whereas the semiarid desert supports mesquite, piñon and juniper, and short growth-cycle annuals and various grasses.

Climatic patterns vary with the elevation. Table 2.2 shows the way mean annual rainfall increases from the "true" desert at Yuma near sea level to the "subhumid" land of Flagstaff at about 7000 feet elevation. The average length of growing season decreases with elevation. The Imperial Valley in California has one of the lowest annual rainfall records of any inhabited area in the United States. The long-time average is similar to the Yuma desert, about three

Table 2.2

Influence of Elevation on Average Length of Growing Season in the Desert Southwest

Weather station	Elevation	Rainfall (long-term average)	Average length of growing season	Climate
	feet	inches	days	relative terms
Brawley, Calif.	— 128	2.1	365	True desert
Yuma, Ariz.	138	3.0	364	True desert
Sentinel, Ariz.	690	4.4	307	True desert
Phoenix, Ariz.	1,114	7.2	294	Desert
Tucson, Ariz.	2,526	11.0	265	Desert
Holbrook, Ariz.	5,069	8.9	174	Cold desert
Flagstaff, Ariz.	6,993	18.3	120	Subhumid

inches per year. Southeastern California, like southern Arizona and southwestern New Mexico, is in a climatic zone of maximum sunshine in the United States. Climatic conditions of southern Nevada are similar to those of the Sonoran Desert of the Southwest.

Physiography of the desert Southwest is characterized by elongated mountain ranges rising abruptly from broad plains or basins. A change in climate dramatically follows the rapid changes in elevation. The plains or basins are filled to varying depths with heterogeneous geologic debris partly eroded from adjacent mountain ranges and partly brought in by streams and wind. The thickness of the deposits may extend for several thousand feet. The coarseness varies. Alluvial (water-deposited) fans flanking many basins in California and Nevada, for example, are composed of coarse material overlying soft, fine-grained deposits. Similar relationships of texture (sand, silt, clay) exist in Arizona and New Mexico.

Diseases, Weeds, Insects, Animals

The harsh climate of the desert leads many people to believe desert soils are sterile. This is not so. Even the sun-baked surface contains millions of bacteria, fungi, algae, and other microorganisms per gram weight (454 grams equal one pound). Only a few may infect plants. About the same kinds of microorganisms inhabit all soils of the world regardless of geographic location. Soil-borne microorganisms which cause diseases in plants (plant pathogens) are not as widespread as non-disease producing organisms (non-pathogens), since pathogens depend on the presence of the host plant species. Growing the same plant in a soil over a period of time may build up plant pathogens, fouling the soil for that specific plant grown. This can happen in desert soil as in any other.

Soils do not respond to inoculation with *microorganisms.* The resident new to the desert frequently is bombarded with sales talk urging him to inoculate the soil before landscaping or planting a garden. Anyone who succumbs to this practice is wasting money and may indeed be fouling the soil with contaminants. The indigenous microorganisms usually eliminate or crowd out the newly introduced organisms, and the native microbial ecology is maintained. One kind of inoculation, and one only, has proven useful to the grower, and that is concerned with legume plants (peas, beans, sweetpeas, alfalfa, sweet clover, etc.). The legume bacteria (*Rhizobia*) invade roots of a specific host species and supply them with available nitrogen. Many leguminous plants grow in the desert. Palo verde, acacia, mesquite, bur-clover, lupine, desert ironwood, and wild vetch are a few.

Numerous *insects* inhabit the desert. A large number are nocturnal. Many of these may be classed as "desert-evaders," since they live in the cool depths of the soil seeking protection from the heat during the day and feed during the night. A second observation is that some diurnal insects have special adaptations to reflect the strong sunlight. Most of the insects, some of which are provided with venom, are limited to the desert rather than cooler, damper climates surrounding the desert. As new plants are introduced into the desert, new insects appear. Often those already present increase in population if the new plant suits their feeding habit.

Animals other than insects can influence the way plants grow in the desert. Pack rats, kangaroo rats, mice, ground squirrels, and rabbits redistribute seeds and attack cactus and succulent annuals and even eat the bark of shrubs and trees. Rabbits make gardening difficult in suburban areas. Their appetite for vegetable garden plants and ornamentals, as well as for native shrubs and trees, is unlimited. Land-living tortoises and lizards, represented by the chuckwalla, are plant eaters. The desert tortoise and numerous toads appear as if from nowhere during summer thunderstorms and feed on native vegetation and insects.

Among the most interesting and well-known desert invaders are tarantula, centipede, and large palo verde root beetle. Despite regular periods of drought, hot dry weather and low humidity, they appear prevalently in desert soils. Their burrowing action mixes and transports organic matter in the upper horizons of soils. Their tunneling over centuries has left avenues for salt, gypsum, and lime to deposit.

Weeds rob the soil of moisture and nutrients which would otherwise be available for desired plants. The weed problem adds to the cost of maintaining landscapes and home gardens. Weeds also discourage the homeowner's interest in improving and beautifying his property. Some range weeds poison animals. The greatest problem, however, is associated with thorns, barbs, stiff hairs, and sharp-pointed seeds that cause problems with man as well as pets. In fact, serious mechanical injury may involve eyes, intestines, and skin and hide of animals.

What is a weed? A weed is any plant that grows out of place or is unwanted because of certain undesirable features. A weed may be desirable in one environment and undesirable in another.

Weeds in your garden, the lawn, alleys, and vacant lots do not all originate from the surrounding desert. Some come from outside of the United States; for example, Russian thistle, Johnson grass, London rocket, puncture vine, knotweed, field bindweed, and wild oats. Weeds such as alkali heliotrope, sunflower, sandbur, and crownbeard have been brought in from other parts of the United States. Other weeds, such as sprangletop, blueweed, slimleaf bursage, careless weed, and white horsenettle, come from the surrounding desert. Thus plant ecology in residential areas and farmland in particular has been altered by man's moving to the desert. Weeds compete with more useful plants for precious water and soil nutrients.

Soil Factors

The most distinguishing characteristic of desert soils of the Southwest, other than droughtiness, is their great variability, which extends to all qualities. Their texture ranges from the extremes of sand to clay, and from stony foothills to colloidal (fine clay) playas. Some accumulate salt; others are free of salt. Some contain soluble elements toxic to plants; others none. Lime is present in most soils, yet a few are free of lime in the surface. Lime is thick-layered, thin-veined, nodular, or soft powder. Some soils require reclamation treatment, and others produce unusually high-quality vegetables, flowers, and landscape plants without special treatment. Shallow and deep soils either limit or favor deep-rooted plants.

The complexity of soils, both in composition and physical condition, makes it all the more necessary to know what they are like and how they react to tillage and irrigation treatment as a medium for plant growth. The average desert soil by volume is nearly 50 percent mineral matter, with a small percentage of organic matter. The remaining 50 percent volume is pore space, which contains air and water. A balance between air and water in the pore space is all-important. When water is excessively high, waterlogging occurs and plants rot. When air excessively dominates, soils are droughty and plants wilt and die. A gardener must know how to maintain a practical balance between the two extremes or he becomes a failure in his own backyard. He must know, for example, that the water-holding capacity of a fine-textured soil (clay) is greater than that of a coarse-textured soil (sand). Furthermore, different types of soil structure favor rapid infiltration rate and deep water penetration, whereas others do not. To make all this more complicated, different plants require different amounts of water, and all water is not absorbed equally well because of differences in the salt content of soil water.

Effects of Man

Plants in the desert grow in a particularly delicate balance with their habitat. The ecological balance is precarious and extremely sensitive to abuse. The solution to living in the desert is not confined to water, for water alone may cause other problems while solving the moisture deficit. Water management is complex. *Mis*management of water results in saline and alkaline soil conditions, putting land out of productivity. Even the best water management practices are not satisfactory without soil and plant management. Irrigation water disposal is an international controversy and a private grower's dilemma. Lush oases in a year round growing climate invite rapid and intensive invasion of plants by insects, diseases, and weeds. Grasshopper, cricket, and locust invasions testify to man's interference with the natural biological balance in arid climates.

Overgrazing of desert lands changes the ecological balance of range plants. A shift to plants less desirable for animal feed has reduced the capacity of the land for grazing and established a high population of noxious plants. Some lands have had to be abandoned because of surface erosion and scarcity of grass. Destruction of the original natural vegetation encourages excessive soil and water erosion, further downgrading the soil fertility and the soil's productive capacity. Often costly flood control and water spreading structures are required.

Dust storms during periods of strong winds associated with droughty or dry soil too often have been the price paid for mismanagement of soils. Fatalities to automobile drivers and passengers have occurred as a result of blinding dust blown from dry cultivated fields adjacent to highways.

These problems of a large-scale nature can become a multitude of small-scale problems for homeowners, park and golf greens superintendents, and municipal landscapers.

Destruction of the desert as a result of man's invasion need not be the rule. Through involvement of a better knowledge of the fragility of the desert ecosystem and the technology of soil and water management, and care for preserving the vegetative cover, man can cease to abuse the native desert beauty and pollute the desert habitat. Soil management plays a key role in the achievement of an abundant life in the desert Southwest.

Optimizing the Environment

Soil Profile Modification*

The maintenance of or improvement in the quality of the southwestern desert for good plant growth depends wholly on the way it is managed.

The volume and quality of the soil reservoir available to plant roots are vitally important to the survival of a plant under arid climatic conditions. Irrigated agriculture in the desert Southwest occupies the best land in the valleys. The soils, for the most part, are

*The soil profile refers to the characteristic features of the soil as seen by a vertical cut through the weathered soil mass into the relatively unweathered geologic material and finally to bedrock (see Chapter 2).

deep and drain well. The opportunity for root penetration is inherently good. Residences and municipal areas, on the other hand, locate on both good and marginal quality agricultural land. Indeed, some of the most valuable homesites are in foothills, footslopes, and on rocky, steep terrain where the soils often are shallow, stony, salty, or limy. These soils require modification for good plant growth. Even with the best soil conditions, physical obstruction to root penetration occurs with continued traffic, cultivation, and other uses: compaction worsens, salt and dispersed clay clog pore space, structure deteriorates, and hard spots develop. A reduction in pore space from 55 to 34 percent has been shown to eliminate plant growth. Traffic compaction is typified by the wheel marks seen in Figure 2.4. Differential infiltration of water leaves some parts of the soil too dry and others too wet. Roots do not extend into dry or into an excessively wet soil.

Fig. 2.4. Inhibition of plant growth over tractor wheel path. Compacted soil under tracks reduces soil pore space.

To provide adequate root environment, one of the prime criteria for growing plants is the selection of soil deep enough to accommodate plants to be grown. If this is not possible, then obstructions, compaction, hardpans, or textural stratification should be modified (Figs. 2.5 and 2.6). Shallow soils over caliche can be improved, depending on the nature of the caliche. Soft caliche that is nodular, veined, or thin-layered may be removed by deep spading and raking

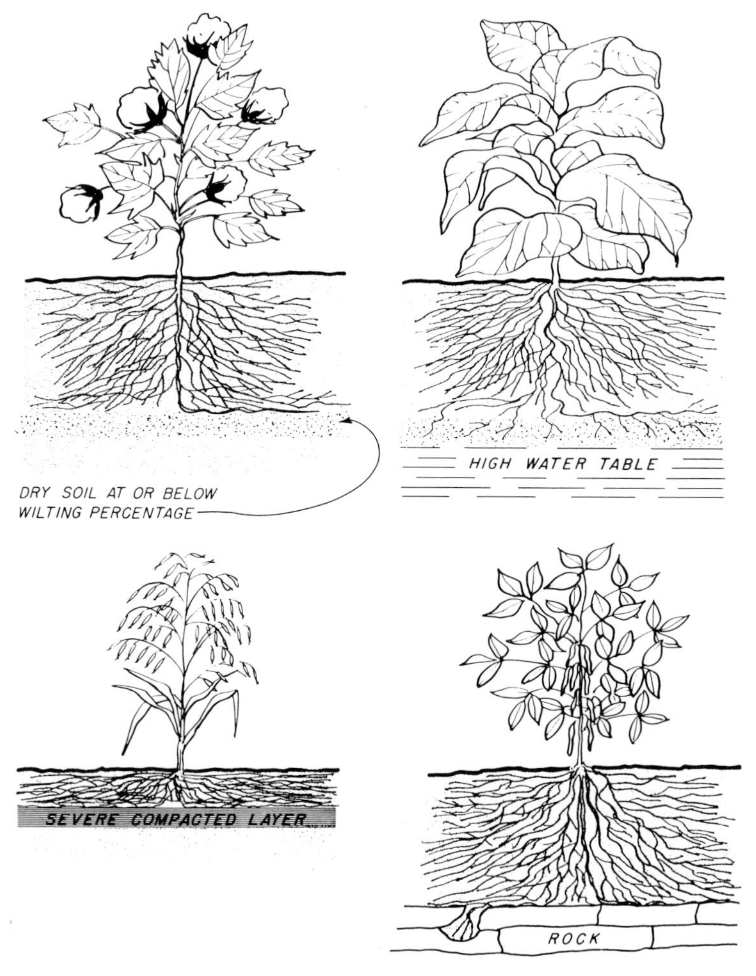

Fig. 2.5. Four common conditions — dry soil, high water table, compaction, and shallow rock or caliche obstruction which limit root feeding area (from US SCS *Engineering Handbook,* Sec. 15, *Irrigation* Ch. 1, *Soil-plant-water relationship.* U.S. Govt. Printing Office, Washington, D.C., 1964).

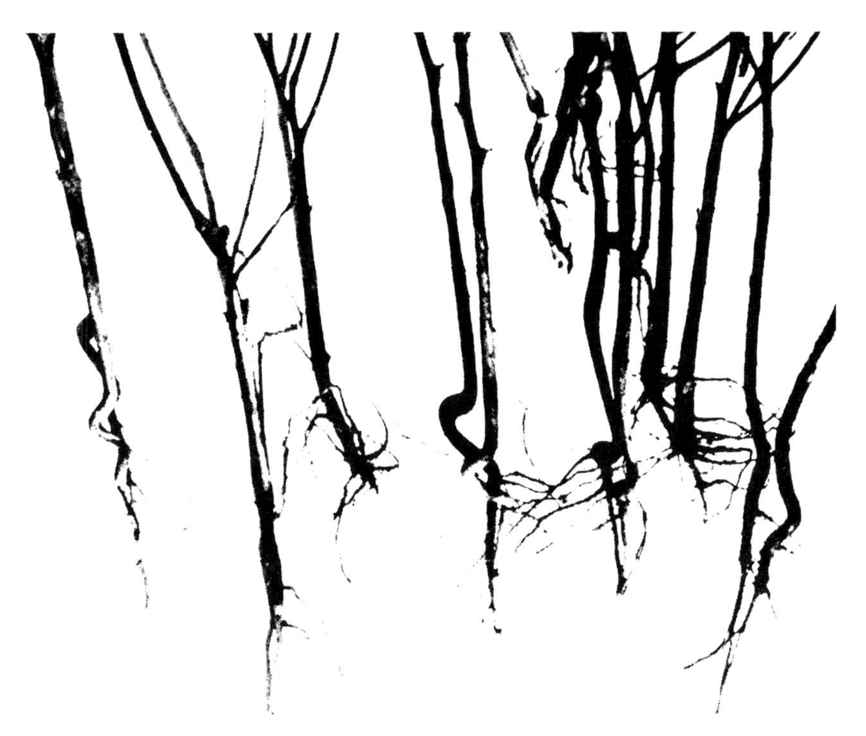

Fig. 2.6. Plant roots, twisted and short. The extreme limited growth is due to compacted soil near the surface.

away the lime. The soil of flower beds and tree and shrub holes may be removed, stacked, and sieved to eliminate stones and caliche before planting. If the caliche is dense or rocklike, it is discarded with the associated limy soil in favor of newly acquired soil. Thick layers of porous caliche are removed also in the same manner. It helps to soak the proposed water basin or tree hole with water a few days to soften the caliche before excavation. Tree and shrub holes are dug either through the caliche, or into it, and a hole punched through it into the loose sand usually present below. A pick and crowbar are standard gardening equipment for many homes in arid lands. Field crops must tolerate the caliche, since the expense of removal is prohibitive to economic production. Ripping or blading caliche after deep irrigation helps to improve root penetration. Shallow-rooted plants grow over caliche.

The bad effects of textural variations or stratifications in the absence of caliche can be lessened by deep spading or some treatment involving deep soil mixing. Water penetrates poorly where textural stratifications of sand, silt, or clay occur.

Texture and Structure Improvement

Two important factors affecting root environment quality are *texture* (sand, silt, and clay) and *structure* (the arrangement of sand, silt, and clay into structures, as crumb, blocky, etc.). Sandy soils become droughty and clay soils compact, and therefore water infiltrates very slowly. A desirable texture lies somewhere between these extremes. Texture may be improved with difficulty and at high cost. Home gardens and landscapings that have texture problems commonly depend on soil brought in from better sources. Topdressing of turf seedbeds with several inches of red mesa soil is often practiced in the larger cities. Addition of steer manure, compost, and plant residues such as straw or sawdust provides some correction of soils with texture or structure problems.

The arrangement of the individual soil particles in *structures* as illustrated in Figure 2.7 controls water movement in soils. Soil structure can be improved almost always by addition of *organic material.* Large quantities applied during the early phases of correction seem to be required in desert soils. Layers of steer manure, up to six inches, are applied to seedbeds for lawns. The organic material is worked into the upper foot of soil or layered six to ten inches below the surface. Most landscapers prefer the mixing of organic material with the soil for best results. Materials such as compost, grass clippings, peat moss, bark material, and pumice also improve the physical properties of the root environment when mixed into the soil.

Irrigation

More water, of course, is needed for home gardens and lawns in the desert than is available from scant rains.

Management of soil goes hand in hand with management of water. The interrelationship is so close that scientists often refer to management of irrigated land as soil-water management, linking the two resources into a single compatible concept.

The most frequent question asked by homeowners, particularly those who come from a different climatic region, is, "When should I irrigate?" Others are, "How much should be applied?" and, "How will I know there is enough water?" No single answer satisfies all soils, all water qualities, or all plants. Watering plants and irrigating

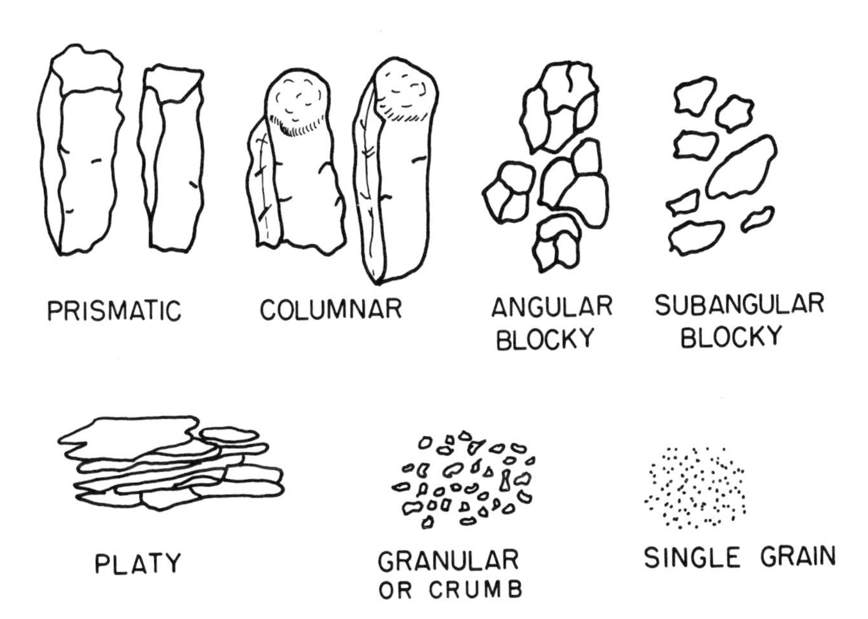

Fig. 2.7. Various types of soil structures found in desert soils.

crops still requires some art as well as science, i.e., one must have some feeling for plant life. However, with a knowledge of soil, water, and plant characteristics and an understanding of basic principles of the interrelationships between them, fairly accurate predictions can be made in answer to these critically important questions. Some of the characteristics that affect water use for irrigation are: salt content, toxic elements, proportion of sodium to calcium and magnesium, amount available, quality-consistency throughout the year, and nitrate content.

When plants wilt, it is evident of course that water is needed. Allowing plants to wilt is drastic treatment. Relying on this method for knowing when to water results in poor growth and poor quality plants. Water must be kept available to plants at all times just as it must for man. The importance of water to plants can be realized in part by the fact that *500 to 1000 pounds pass through plants growing in a humid climate for every pound of dry matter produced.* In the desert the high rate of transpiration makes this demand even higher. High evaporation from the soil surface adds to the water demand. Several inches of water in tree and shrub basins once a week does not seem excessive during the spring, summer, and fall months. The best method is to dig into the soil with a probe or trowel. The soil should be moist to the feel. The soil should be soaked throughout the root zone thoroughly during irrigation, taking into account the different kinds of root systems (Fig. 2.8).

An abundance of water is critical during two periods in a plant's growth cycle. One is during seed germination and seedling establishment. The soil should never dry out in this stage of plant growth. The second most important growth stage is when the plant is larger and growing at an accelerated rate. Larger amounts of water and deeper penetration are required then than earlier in the seedling stage. Most garden plants grow well in soil wet to three feet. Clay

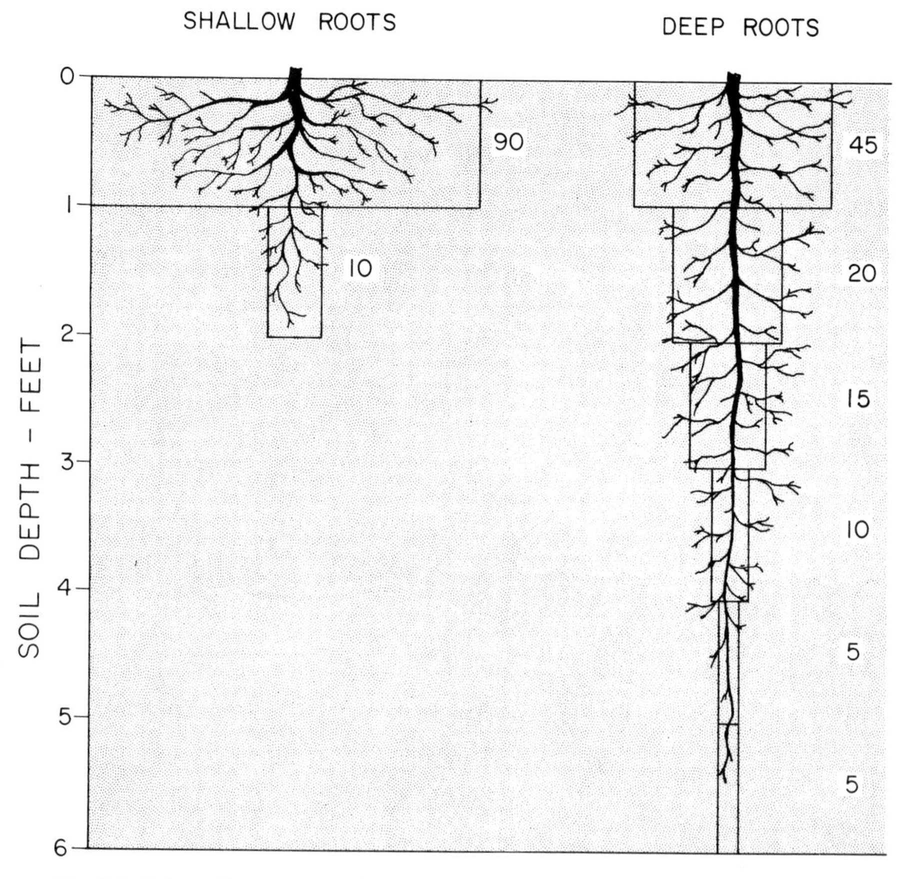

Fig. 2.8. Schematic representation of two root patterns and their percentage use of water at various soil depths (data from R. L. Hausenbuiller, 1972. *Soil principles and practices.* Wm. C. Brown Publishers, Dubuque, Iowa).

(or fine-textured) soils require about twice as much water to saturate to a given depth as sands. About one inch of water wets one foot depth of sand.

Fertility

Fertility is concerned with nutrient-supplying capacities of soils. Fertility is defined as the potential of a soil to provide nutrient elements in amounts, forms, and proportions sufficient for maximum plant growth. The concept of soil fertility differs from that of productivity, because it is limited to plant-nutrient supply and does not connote the presence of salt excesses or specific toxic ions, nor physical barriers within the root zone, nor the ill effects of man and animal invasion. On the other hand, productive capacity does include these items. Productivity "counts the sugar beets in the bag," fertility pertains only to the nutrients which are needed to grow that many (or more) beets. Nevertheless, successful growers give great attention to soil fertility because it is one of the growth factors over which man usually has control.

The essential plant nutrients (Table 2.1) comprise the major components of a fertile soil. The elements occur as *simple* and *complex salts*, as *exchangeable elements* (ions) adsorbed on *clay minerals*, fixed in *primary minerals*, and as constituents of *organic matter*. The soluble forms are readily taken up by roots. Less soluble forms become available to roots slowly and more uniformly over a longer period of time. The *rate* at which the nutrient element becomes soluble is as important as the *amount* available at a given time. The plant does not suffer deficiencies, provided the supply rate is maintained at the rate the plant demands.

Nutrient elements have been divided into two general groups as shown in Table 2.1, *macronutrients* and *micronutrients*. Macronutrients are required by the plant in larger quantities than micronutrients. Micronutrients (also referred to as "trace," "minor") are required by the plant in only small amounts. They concentrate in enzymes, vitamins, and other biochemical and growth controlling substances, as well as certain cellular structures.

How can we determine the fertility status of a soil? Home gardeners look at the thriftiness of the plant. A farmer does the same, but also measures yield as correlated to fertilizer-treated and untreated soil. The best way is the use of one or more of the following methods: (1) *plant growth response* — apply the element(s) thought to be deficient to only a small part of the field, then compare growth with those plants not receiving the nutrient(s), (2) *soil analysis* — have the soil tested where correlations between plant response to fertilized and unfertilized fields can be related to chemical analyses, and (3) *plant analysis* — measure growth response and/or element concentration in a specific plant part. Methods (2) and (3) require the services of a reliable testing laboratory, and experienced specialists to interpret results. The County Agricultural Agent or Farm Adviser can furnish information about this kind of service.

Certain soil characteristics, such as black color, which is indicative of an abundance of soil organic matter, are considered to reflect fertility. Since soils of arid regions are low in organic matter, the dark coloration loses much of its significance for prediction. Soils of the desert Southwest over many years have accumulated plant nutrients in their surface layers despite the scanty vegetation and low levels of organic matter. Furthermore, low rainfall has not washed them out of the soil by excessive leaching. Except for nitrogen, phosphorus, iron, and zinc (and sometimes potassium) deficits, arid soils of the desert Southwest are fertile; indeed, excesses often are more common than deficiencies.

Soluble Salts

Soluble salts in the root environment can become excessive for plant growth unless leaching (washing) is practiced to move them below the root zone. Soils that have accumulated salts require reclamation methods such as leaching with good water and/or the addition of gypsum (Fig. 2.9).

Insects and Rodents

Insects and rodents deteriorate root environment. Ants are particularly destructive and require the use of insecticide. Rodents tunnel and burrow, allowing soil to erode and water to channel and waste away.

Fig. 2.9. Accumulation of salt on the high centers of lettuce beds (from Fuller, W. H. 1962. Reclamation of saline and alkali soils. *Plant Food Review* 8[3]: 7–9).

Diseases

Some soils may harbor root diseases which result in "damp-off" of seedlings. Growing the same plants in a specific area year after year may cause build-up of certain disease organisms. Rotation of different kinds of plants is suggested to remedy this problem. Soil fumigation may be used with good success.

3. Soils Have Profiles

Leaf after leaf of a book of soil profiles unfolds as we drive along our modern highways and roads. Fresh cuts often deeply slice through the soft mantle of the soil body down into geological parent material and the solid bedrock below. The observing traveler encounters a wide variety of different soils and cannot help but notice their dramatic variations in color, texture, depth, and stratified material. Judging from numerous requests (received by soil scientists at the University of Arizona) for samples of desert soils by students working on science projects all over the United States, there is an increasing awareness that soils differ in different locations. Indeed they do. Even within a small area of a few acres, soils differ in appearance. Wide differences occur in fertility and productivity also. Soil-forming factors (climate, vegetation, parent material, topography, and time) impose on numerous geological parent materials making soils differ in the way they look and support plant growth. Let us examine some characteristics of southwestern desert soils.

The Profile

A vertical cut through the soil body down into the geologic material or rock from which it originates exposes the soil profile (Fig. 3.1). The characteristics of each soil reflect the combined soil-forming forces working over a long period of time to form different layers or *horizons*. The sum of these horizons is called the *soil profile*. The soil profile as described by the various horizons is what appears so clearly in fresh road cuts. The soil profile informs us about the root feeding zone: its depth; the organic matter content; salt accumulations; texture; structure; and many other characteristics important to the welfare of the life it supports. Soil profiles differ because no two horizons are identical. Not only do they differ in proportions of the above constituents, but they differ in plant nutrient levels and

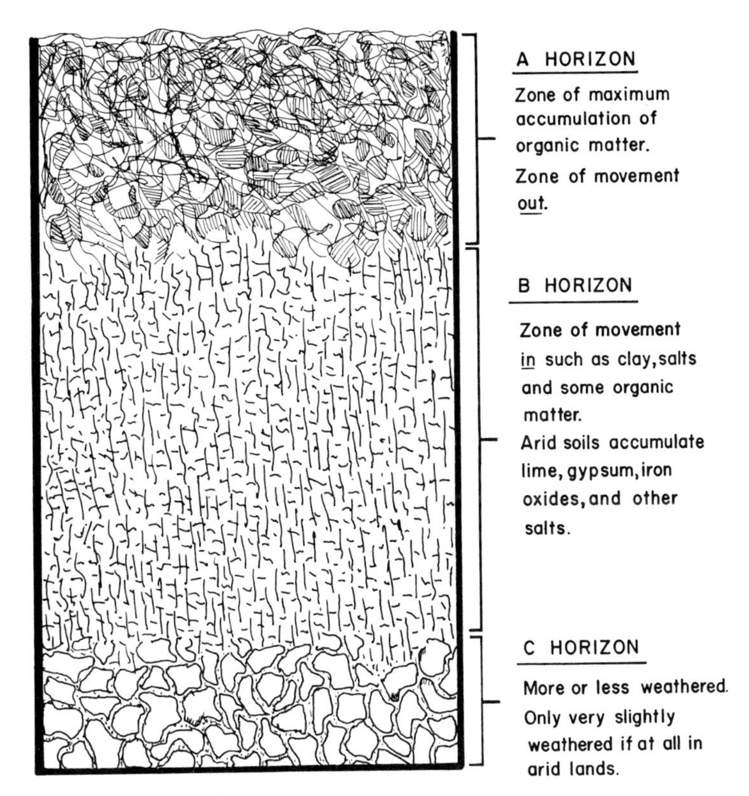

Fig. 3.1. Schematic diagram of a soil profile showing the horizons in which plants root most extensively (from *Soil management: Humid vs. arid areas,* Thomas C. Tucker and Wallace H. Fuller. In *Food, fiber and the arid lands.* William B. McGinnies, Bram J. Goldman, and Patricia Paylore, Eds. Univ. Ariz. Press, Tucson, 1971).

availability of nutrients to plants. Even in desert soils where the soil-forming forces are slowed by the lack of rain, soil profiles have two or more of the following major horizons:

- O Surface horizon containing mostly decomposing plant debris (leaves, twigs, roots). Usually it is very thin or absent in desert soils.
- A Horizon from which constituents move out (or down) in the soil solution or in suspension as very fine particles; this also contains some organic matter.
- B Horizon of accumulation of materials and substances from the above horizons.
- C Horizon of parent material. In desert soils this undergoes unpredictable alterations.

These horizons, except O, are illustrated in a typical Mohave loam profile diagrammed in Figure 3.2.

The profile of Laveen clay loam is an example of a limy desert soil which is quite productive (Fig. 3.3). Parti clay loam illustrates a fairly productive arid land soil characterized by a shallow profile over a thick layer of lime (Fig. 3.4). The rainfall where Parti clay loam occurs is higher than that of a true desert, and therefore the soil represents a transition type typical of those associated with both the true desert and subhumid climates.

Color

Soil color tells the studious observer much about the soil. In the early history of agricultural development in the United States, land was commonly bargained for on the basis of its dark color. Growers were aware of the association between organic matter, which gives soils of humid climates a dark color, and fertility. The color of soil differs when wet and dry. Wet soils are darker. Mois-

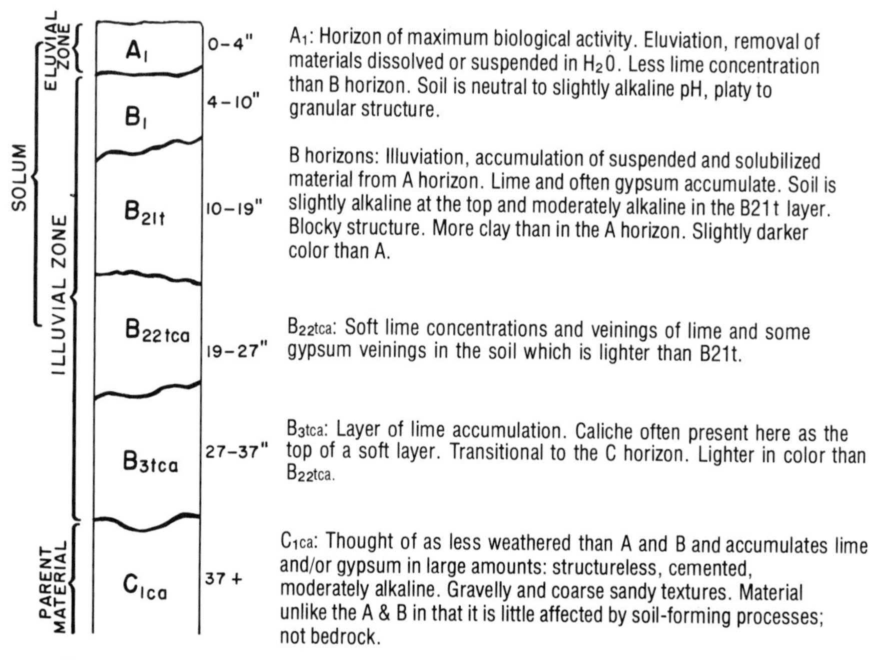

Fig. 3.2. Diagram of a typical Mohave loam profile from the southwestern desert.

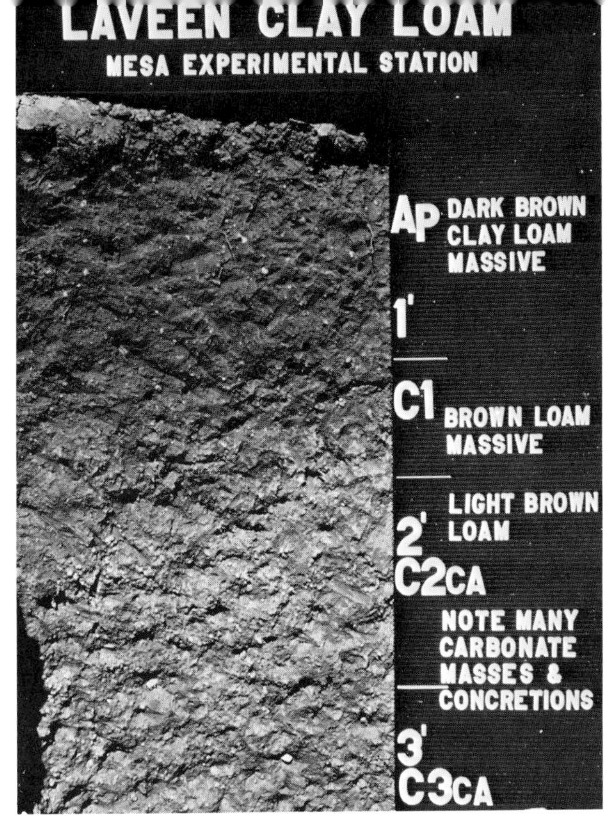

Fig. 3.3. A soil profile of Laveen clay loam from Mesa, Arizona, showing lime nodulation throughout and the highly limy subsoil. The B horizon is absent in this profile.

ture relationships in soils, therefore, may be detected by differences in color. Deep gray subsoils with flecks of iron stains often relate to soggy, wet conditions or waterlogged (overwatered) soils.

Desert soil surfaces are lighter in color than soils of humid regions. This is due to the low levels of soil organic matter and to the usual presence of lime and salts. Red and reddish browns often dominate in the southwestern desert. In fact, landscape architects, golf greens managers, and builders commonly request "red mesa" soil for fill in lawns and gardens. The gray river-bottom soil contains larger concentrations of salt and alkali. It is less desirable for plantings than the red mesa soil.

Elaborate color charts have been developed to help identify different soil horizons for soil classification purposes and study. In fact, color is so important for identification that two great North American land divisions carry the common names Gray Desert and Red Desert, the gray color being associated with the colder desert of the Great Basin area of the western United States.

Fig. 3.4. A Parti loam profile. This soil has well developed A, B, and C horizons. The clayey B horizon is located as the area within the limits of the geologic pick. Note the very limy C horizon which is characteristic of many desert soils.

Organic Matter

Though organic matter is low in soils of the southwestern desert compared with soils of humid regions, the small amount of organic matter is vitally important to plant growth. It exerts a profound influence on soil structure, water movement and retention, and aeration. The maintenance of this relatively small amount of organic matter in southwestern desert soils is essential to successful gardening and beautiful landscape establishment. Usually southwestern desert soils do not exceed 0.1 to 1.0 percent organic matter. Four to five times this amount is common in midwestern soils.

Soil organic matter is one of our natural resources that has been exploited and dissipated by modern civilization. Yet civilization's waste materials for suitably maintaining the native level are lying idle or polluting our land and water systems. The invisible decline in soil productivity by intensive cultivation and removal of plant debris, which often could be returned to the soil, is as equally detrimental to commercial crop growing and to beautification of residences as any other manageable factor. Continued ability of the soil to support favorable plant growth will depend on how well we manage the organic matter program and return suitable plant residues to the land.

Soil organic matter in southwestern desert soils contains a higher proportion of both nitrogen and phosphorus than that of humid climates. The full value of soil organic matter is difficult to assess. It is known to (1) provide plant nutrients; (2) improve the physical properties of soil and thereby control soil moisture relationships; (3) supply food for the soil organisms which benefit plant growth in many ways; and (4) assist in control of pathogens of plants, animals, and man. Almost all of the antibiotics so important to the health of man are obtained from soil microorganisms which depend on soil organic matter for food.

The rapid rate of decomposition and recycling of residues in southwestern desert soils is phenomenal. Large quantities of residues disappear from irrigated land in a relatively short time because of the warm temperatures throughout the year.

Texture and Structure

Texture. Texture refers to the individual particle sizes as sand, silt, and clay (Table 3.1). When one size dominates, the soil is called *sandy*, *silty*, or *clayey*, depending on the particle size. The

Table 3.1

Soil Particle Size

Soil particle class	Diameter range	
	Millimeters*	Inches
Very coarse sand	2.0 –1.0	0.0787–0.0394
Coarse sand	1.0 –0.5	0.0394–0.0197
Medium sand	0.5 –0.25	0.0197–0.0099
Fine sand	0.25–0.10	0.0099–0.0039
Very fine sand	0.10–0.05	0.0039–0.0020
Silt	0.05–0.002	0.0020–0.0008
Clay	Less than 0.002	Less than 0.0008

*1 millimeter = 0.03937 inch

Table 3.2

Percentages of Sand, Silt, and Clay in Some Common Soil Classes

Soil class	Sand	Silt	Clay
Clay	Less than 45	Less than 40	More than 40
Clay loam	20–45	15–50	25–40
Silt loam	Less than 50	50–90	Less than 30
Loams	25–50	30–50	5–25
Sandy loam	45–85	Less than 50	Less than 20

nomenclature also applies to *gravelly* and *stony* soils high in these textural sizes. *Loams* are considered to have an ideal texture for gardening. Loams contain about equal proportions of silt and sand and less than about 25 percent clay (Table 3.2). Soils having more than about 25 percent clay are very sticky when wet, whereas those having more than about 70 percent sand are droughty because they store water so poorly. Fine-textured soils (clays) contain more pore space than coarse-textured soils (sands). Clays, therefore, can store more water than sands. A combination of textures gives a *soil texture class* (Table 3.2).

Structure. Structure and texture of a soil are not the same. Whereas texture refers to the proportion of sand, silt, and clay (individual particles) in a soil, structure refers to the *arrangement* of the particles into aggregates. *Ped* is a name used to describe a natural aggregate. A *clod* is an aggregate formed when soils are spaded, plowed, or worked when too wet. Clods form more as a result of man's activity than by natural circumstances. They are not water-stable, that is, they disperse when irrigated and lose their original form. Some scientists use the term *concretions* for accumulation of certain salts into structures visible to the eye.

Five general soil structure types are recognized: (1) *platy*: peds exhibit a flat, platelike appearance; (2) *prismatic*: peds which have one axis (length) long with flattened sides . . . prismlike; (3) *blocky*: peds appear as cubes, like ice cubes: some larger and some smaller; (4) *subangular blocky*: similar to blocky but more rounded, differing in size from shot to marble; (5) *crumb*: peds are angular and small like bread crumbs. Diagrams of these types appear in Figure 2.8.

Crumb is a structure term often used because it so characteristically forms in grass sods. It is very desirable in a garden soil. Crumb structure is ideal for plant root extension. Air and water move favorably in soils having such a structure.

Some scientists refer to sand as *single-grained* since it is not aggregated. Sandy soils contain many large pores allowing ample to excessive air and water movement. Because water retention in sands is low, sands are considered droughty. They require frequent irrigation.

Puddled soil structure occurs under a circumstance similar to that causing *cloddy* soil. Soils that are worked too wet lose their structure and disperse into single particles. Such soils form a hard-baked surface when dry. Large cracks often break the soil up into lumps and clods. Fine-textured soil such as clays "puddle" more readily than sandy soils when tilled too wet (Fig. 3.5).

Unlike texture, which cannot be readily changed, structure may be modified. Organic residues, composts, manures, sludges, sawdust, and certain other *soil amendments* — gypsum and sulfur, particularly — can improve the structure in soils.

Fig. 3.5. A puddled clay loam soil. Plants can emerge only through cracks. This soil was tilled when too wet and was later dispersed by a heavy summer rain.

Stratified Materials

Soils of the southwestern desert may contain horizontally stratified materials (layers) as a general characteristic of the profile which do not relate to the genetically formed A, B and C horizons. Almost all valley areas show textural stratified effects of flood action. Layers of different textures (sand, silt, clay, and even stone and gravel) fall abruptly upon one another (Fig. 3.6). Arid land weathering processes are so mild they have had little influence on modifying the original textural layering in the profile. Homeowners located in the foothills, footslopes, and alluvial fan positions find

Fig. 3.6. Vinton loamy fine sand may be found in the alluvial material adjacent to major stream beds. Soil development is restricted to an A-C profile with pronounced stratified layers of sands and silts of varying particle sizes and thicknesses.

Fig. 3.7. Layered lake and sea deposits overlay a buried soil profile perhaps millions of years old and formed when land was experiencing a humid climatic condition. Near Las Cruces, New Mexico.

highly layered materials, including gravel and stone. Caliche or lime layers commonly occur as thin and thick deposits, nodular and veined, or as soft, powdery lenses. Caliche can occur in depths from a few millimeters to several meters thick. Gypsum salt concentrations appear in scattered locations. Old buried soil profiles or their remnants containing clayey layers, developed when the desert was under a more humid climate or when the land was covered with lake or sea water (Fig. 3.7), often offer the home gardener and landscape planner real problems. The water-laid clay layer can be so dense that the soils become cloddy when dry and sticky when wet. Water movement is difficult to control, and roots penetrate the clay poorly or not at all.

Not the least observed stratification is that of sand dune formation. Windblown sand may cover any type of soil or rock debris to various depths. Fortunately, wind movement in the Southwest is not as severe as in deserts in other parts of the world. Stabilization of sand can be accomplished in residential and municipal areas. Unprotected desert land, however, still has problems of productive soil being covered with creeping sand dunes.

Salts

Salts can concentrate in soils at any level. They may stratify on the surface by capillary movement upward, due to improper soil and water management, or they may stratify under some natural condition. Where textures change abruptly salts may concentrate in the clay lenses or layers. Leaching the salts below the root zone is the only real solution to ridding the soil of salts harmful to plant growth. Salts may accumulate sufficiently to prevent plant growth, as seen in Figure 3.8.

Fig. 3.8. Salt accumulations have prevented plant establishment.

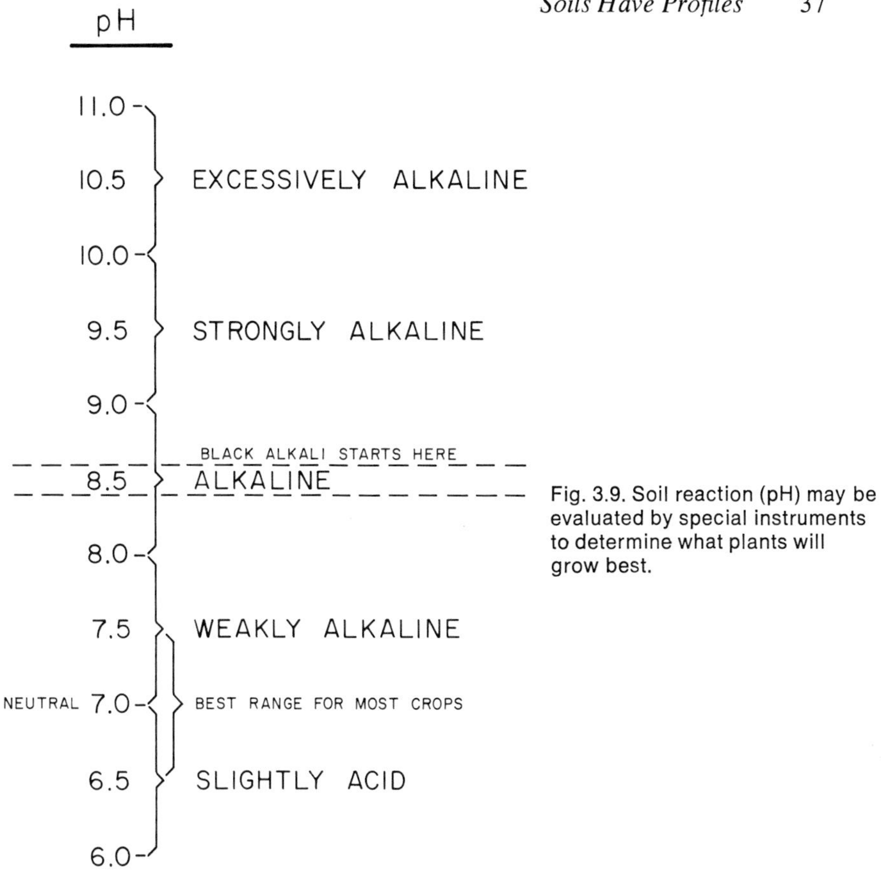

Fig. 3.9. Soil reaction (pH) may be evaluated by special instruments to determine what plants will grow best.

Soil Reaction

Soil reaction refers to the acid (sour), neutral, or alkaline (caustic) condition of the soil. Desert soils notoriously react alkaline to tests. In fact, some named locations refer to the alkaline condition of the soil (Alkali Flat, Alkali Gulch, Alkali Junction, etc.).

Soils of the desert Southwest, fortunately, are not too excessively alkaline in reaction for many plants to grow well. Acid-loving plants, though, do poorly in desert soils. They should not be grown unless some permanent provision is made to acidify the soil and keep it on the acid side of neutral pH 7.0 (Fig. 3.9). Most of the arid and semiarid soils of the southwestern desert have a pH value below 8.2 and above 6.8, which is well within the growth tolerance of a host of ornamental plants and fruit and nut trees.

4. Why Organic Matter?

Most people realize that organic matter benefits soil and that dark-colored surfaces are more desirable than light-colored surfaces because they contain more organic matter. In the desert Southwest, valley soils do not have as dark a surface as soils in humid climates. Only in a few of the upland soils, where rainfall is higher, are soil surfaces dark (though under pine forest conditions the organic layer still is thin). Some of the best desert soils have red, reddish-brown, and gray-brown surfaces. Organic matter is low in these soils. Rarely does it exceed one percent by weight in the upper six inches of the profile.

While the total amount is small in the desert, this small amount exerts a great influence on soil productivity. The maintenance of even small quantities of organic matter in soils is essential to continuing high crop yield, productive home gardens, and beautiful landscapings.

Functions

Organic matter functions as it decomposes. As straw, leaves, sawdust, or animal manure decompose, new synthetic compounds are formed which coat and thread soil particles together into porous structures, enhancing the quality of the physical condition of the soil. Soils with poor structure or none at all become compacted, with accompanying slow water penetration and air movement which are so essential for root development. The channels and crevices of well-structured soil provide necessary avenues for root growth and storage places for air and water, just as building structures provide rooms for human habitation. Breakdown of soil structure is as critical to plant roots as breakdown of building structure is to man.

Soils with good structure are easily worked by implements and are referred to as having good *tilth*. Figures 4.1 and 4.2, comparing

Fig. 4.1. Soil plus manure. This Mohave silt loam received 20 tons of steer manure for five years. The soil structure is loose and granular.

Fig. 4.2. No organic matter. This Mohave silt loam did not receive manure or any other organic residue. The structure is massive and cloddy.

manured and non-manured Mohave soil, reflect the favorable influence of animal wastes on the quality of tilth. Where no organic residue is added, the soil washes badly and has limited pore space. Soils devoid of organic matter (except for loose sands) dry into hard, massive lumps, which provide poor seedbeds and bake into hard surfaces. Further, erosion accelerates and ground water storage becomes inadequate for plant growth.

Mere addition of coarse organic material is generally not sufficient to produce good tilth. Rather it is continual degradation of organic matter by microorganisms which permits good structure to form. Growers for centuries have recognized the benefits of putting organic residues into soils and the desirability of maintaining humus.

Humus is a product of decayed organic materials in the soil. It is a mixture of altered plant and animal residues, resynthesized microbial compounds, and microbial debris. Humus structurally does not resemble the material from which it originated and is composed of finely divided particles which are even finer than clay particles. It functions in providing good soil physical condition.

Organic matter contains valuable *plant nutrients* which the decay processes release to the soil. In fact, all the kinds of nutrients necessary to grow a new cycle of plants are found in the returning residues. The residue, through microbial degradation, supplies nutrients in *mineral form* to succeeding plants. In native soils, nitrogen is carried almost exclusively in the organic matter. As the organic matter decomposes, mineral nitrogen is slowly released. The advantage of nitrogen being originally in organic matter in the form of protein lies in its resistance to loss by leaching. Because of their small size and decay mechanisms, microorganisms have the first chance at using the nutrients released from materials entering the soil. Their population development coincides with the quantity of organic residues supplied to the soil. As the decay proceeds to completion, mineralized nutrients such as phosphate (PO_4), nitrate (NO_3), potash (K_2O), and sulfur (S) accumulate, since they no longer are required by the microorganisms. Plants now have an opportunity to use them.

The organic matter cycle in soils (see Fig. 4.3 and Fig. A.1) illustrates the transformation of organic residues by microorganisms, with the release of simple inorganic compounds as plant nutrients

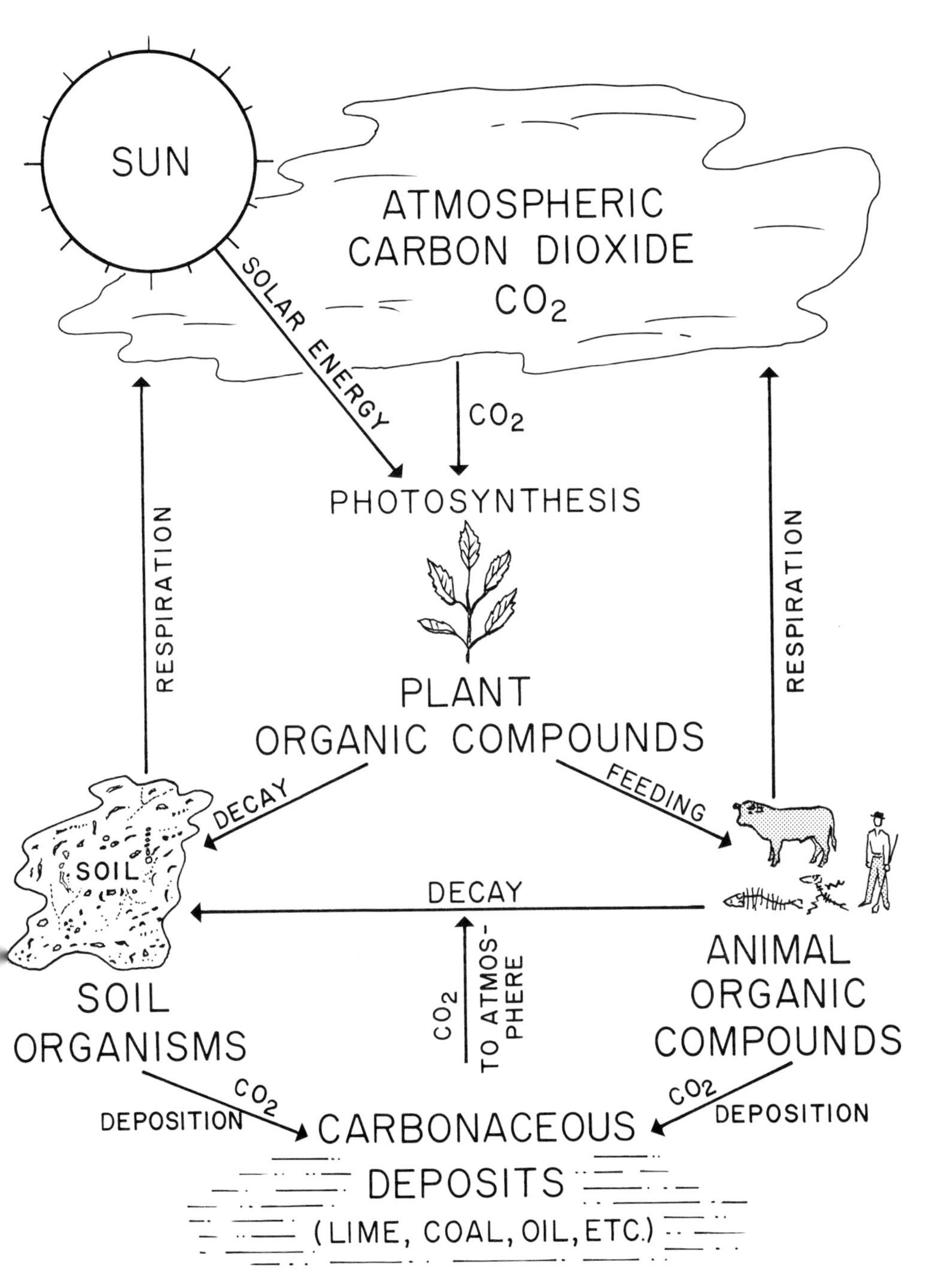

Fig. 4.3. The carbon cycle of the earth.

in a form available to plants. The soil is a pool for storage of these nutrients. The same chemical forms (compounds) for plant use come from degradation of wastes in soil as from commercial fertilizer supplied from a bag. Natural deposits of nitrates in soils (such as the nitrate beds in Chile) originate from fossil organic matter sources and/or accumulation by microorganisms. Even the most resistant plant residues or parts of residues such as lignin eventually succumb to microbial attack, for no substance remains indefinitely in the soil unaltered.

Organic matter functions also indirectly in plant nutrition by hastening the rate of *release of nutrients* from the mineral part of the soil as well as from the organic part. For example, addition of straw or manure to soils has been shown to make the slowly soluble phosphates more readily available for plant use. The degrading organic constituents formed (acid and other functional chemical groups) react chemically, mobilizing the mineral compounds into available (more soluble) forms so they can be absorbed directly by roots. Often micronutrients such as iron and zinc become available, and deficiencies are corrected, when organic plant materials are worked into the soil.

Another important function of organic matter is to act as a source of *food and energy for the maintenance of soil microorganisms.* Life could not exist, man and animal would disappear from the earth, if it were not for microorganisms. The organisms need food and energy to live. Organic matter furnishes the necessary supply of food. The size of the soil population of microorganisms is controlled by the amount of organic matter in the soil as well as the amount of residue returned to the soil. The numbers are so great that many pounds of nitrogen, phosphorus, and other nutrients occupy their tissues in an acre of soil. All kinds of microorganisms inhabit the soils: *viruses* — the smallest living things; *bacteria* — so small they cannot be seen with the naked eye; *amoebas, protozoa, algae* — single-celled animals or plants; *fungi*; and *actinomycetes* or *streptomycetes* — threadlike chains of cells. Bacteria number into the billions per gram (454 grams equal a pound) of soil. Actinomycetes number into the millions, and fungi appear in hundreds of thousands per gram of soil (see Table 4.1). Soils in the desert Southwest frequently contain greater numbers of actinomycetes than bacteria. This suggests that actinomycetes can withstand

Table 4.1

Numbers of Microorganisms Detected in an Irrigated Arizona Desert Soil, Mohave Sandy Clay Loam

Organisms	Number per 1 gram* of dry soil
Bacteria	4,000,000 to 300,000,000
Streptomycetes	1,500,000 to 50,000,000
Fungi	100,000 to 1,000,000
Yeasts	1,000 to 75,000
Protozoa	1,000 to 100,000
Algae	100,000 to 1,000,000

*453.6 grams = one pound

the drouth and high summer heat better than most bacteria. This also could suggest that the actinomycetes can survive better on highly degraded, resistant organic matter that remains in the soil after the initial rapid attack of bacteria.

Algae are important to the soil economy of the desert by their contribution of organic matter and nitrogen. Crusts of algae and lichens inhabit much of the bare desert soil surface. These nitrogen-fixing organisms are abundant, making soil crusts high in nitrogen. Algae appear to be the forerunner of life on a (prehistoric) mineral earth. Many present-day species are identified as being capable of utilizing an all-mineral diet of nutrients from rock and rock debris, and of combining atmospheric nitrogen gas with carbon dioxide to form tissue protein and cells with the aid of photosynthesis or sun energy. They require no combined nitrogen or organic carbon compounds. Algae cling to stones so hot in summer they cannot be handled without gloves (Fig. 4.4). They withstand months of drouth and frost. Within a few hours after a warm summer rain, algae have been found to double their (dry) weight. Algae receive attention here because they grow well on mineral soils and rock and sand debris so much a part of the desert, and in this way act as a necessary forerunner to certain other life in the desert.

Composition

When *lignin* decomposes in the soil, the initial attack of stripping off some of the functional groups (such as carboxy, methoxy, hydroxy, phenolic, ketonic, aldehydric, etc.) is quite rapid, like

Fig. 4.4. Algae clinging to a quartz stone from the desert soil surface.

the decay of the husk of a walnut, leaving the more resistant inner part, the shell. A resistant portion remains that does not resemble the original lignin and degrades slowly to the soil organic pool called *humus*.

The *carbohydrate-like* material in soil organic matter is primarily of microbial origin. Slimes, gums, and organic salts (of uronic, teichoic, muramic, fluvic, and humic acids) contribute to this fraction. These act as cementing agents for good soil structure and react with mineral constituents of soils assisting in nutrient release. The activity and numbers of soil microorganisms have been correlated with the formation of water-stable soil aggregates, or crumb structure. Fungal and algal threads grow through the cracks and crevices of the soil, giving soil structure stability. In this way microorganisms contribute to water behavior in soil by helping to keep the soil porous and friable.

A less desirable aspect of microbial behavior in soil is an excessive fabrication of microbial tissues and threads by certain fungi, which reduces the wettability of the soil, resulting in dry spots. If such microbial activity extends too far, plants can suffer drought and die despite an usually adequate watering program. Excesses of organic residues poorly distributed as lumps or layers can actually act as water barriers when actinomycetes and fungi become overactive.

Soil algae also contribute to poor water penetration by coating pond and waterway bottoms with gelatinous material. At the date of this writing (1974) there is no practical chemical method of eliminating their imperviousness to water. Physically scarifying the soil surface with tillage instruments is suggested to break up the crusts to allow water to infiltrate better. Algae flourish in ponds and lakes and are an expensive nuisance in swimming pools. The warm summer temperatures encourage algal growth. Irrigation canals require frequent dredging to prevent plugging with algae. On the other hand, they act in range stock ponds to seal the soil against downward water seepage. So algae can be pests or benefactors.

A third major component of soil organic matter are the *protein-like materials.* They also originate chiefly from microbial cells and tissues. Irrespective of their source, they become generally modified from their original form. These modified protein compounds are the soil nitrogen carriers. Organic matter of the desert Southwest is richer in nitrogen-containing constituents than that of humid climates, as evidenced by the lower carbon/nitrogen ratio of 8 to 10 compared with 10 to 12 in northern regions.

A fourth component which can be well-identified are the phytin-like *phosphorus* compounds. Organic phosphates accumulate in soil organic matter because of their resistance to rapid microbial degradation. It is believed they originate from both plant phytin and microbially synthesized phytinlike compounds. Organic phosphates contribute slowly to the phosphate economy of soils. They, too, like other complex organic compounds, are not directly available to plant absorption but must be mineralized to release inorganic phosphates before being absorbed by plants.

Sources

Any organic substance if biodegradable (decomposable) can be a potential source of soil organic matter. Sources vary from highly desirable to undesirable. They vary in effectiveness with the purpose for which they are to be used, and the nature of the garden, landscape, or agricultural crop. The cost also varies. The homeowner who needs only small quantities of organic material can afford to pay relatively more than the municipal or industrial landscaper or the farmer.

With the modern focus on utilization of wastes that threaten to pollute man's environment and the emphasis on recycling, new organics appear regularly at the homeowner's shopping centers. Certain organic materials which once were disposal problems are such popular sale items they cannot keep ahead of the demand. Sawdust and bark are examples, as are some animal manures which are helping to fill the nutrient-depletion gap caused by commercial fertilizer shortages.

The following discussion on sources of organic matter will be concerned with natural (or slightly altered) organics. A more complete discussion will appear later where the whole subject is reviewed in detail, and organic compared with inorganic soil conditioners.

Some Home Garden and Landscape Sources

All *animal manures* provide a good source of organic material and nutrients for growing plants (Table 4.2). Their value in the desert Southwest is not always appreciated. The reason for this stems from their low plant nutrient content and high cost of transportation per nutrient weight. A ton of steer manure contains about ten pounds of nitrogen, five pounds of phosphorus (as P_2O_5), and ten pounds of potassium (as K_2O). Secondly, manure from stockyard or pen-fed cattle can be seriously high in potassium and sodium salts (table salt) (Table 4.3), which counteract the beneficial effects of the organic matter in improving the soil structure. Animal manure, free of excessive salt and sodium, benefits plant growth primarily in three ways: (1) It provides an immediate increase in growth and productivity. (2) It provides a long-time improvement on growth and productivity by releasing nutrients slowly. (3) It improves the structure of the soil and its moisture-holding capacity.

Table 4.2

Some Organic Fertilizer Materials

Material	Nitrogen percent	P_2O_5 percent	K_2O percent
Dried blood	8.0–14.0	0.3–1.5	0.5–0.8
Animal tankage	5.0–10.0	3.0–13.0	Small amounts
Garbage tankage	2.0–4.0	1.0–3.0	0.5–1.5
Process tankages	6.5–10.0	Variable	Small amounts
Fish scrap, dried	6.5–10.0	4.0–8.0	Small amounts
Sewage sludge, ordinary	1.6–3.3	1.0	Small amounts
Sewage sludge, activated	4.1–7.5	2.5–4.0	0.75
Cottonseed meal	6.0–9.0	2.0–3.0	1.0–2.0
Bone meals	0.7–5.3	17.0–30.0	...
Castor pomace	4.0–7.0	1.0–1.5	1.0–1.5
Cocoa shell meal	2.5	1.0	2.5
Tobacco stems	1.3–1.6	0.9	4.0–9.0
Sheep or cow manure, dried and pulverized	1.0–2.0	1.0–2.0	2.0–3.0
Poultry manure, dried and pulverized	5.0–6.0	2.0–3.0	1.0–2.0
Bat guano (dry)	1.5–10.0	1.2–4.0	1.0–3.0

Soils become looser and friable and do not bake out or crust as much in the hot sun.

Composted *poultry guano* sporadically appears on the market. It is slightly richer in plant nutrients than steer manure. When guano has been composted and physically processed properly, it loses its unpleasant odor. Some producers fortify animal composts with small amounts of ammonium or urea fertilizers to raise the nitrogen-supplying value.

When stockyard or feedlot manures are used, about ten tons/acre for agricultural purposes, or about one-half pound (on a dry basis) per square foot for home gardens, are recommended. *Animal residues* should be applied at least once a year and well-mixed into at least the top six inches of the soil for best results. Lesser amounts have marginal value on most soils in the desert Southwest because decomposition is so rapid. Leaching of animal salts below the root zone is necessary to prevent undesirable accumulations.

Municipal sludges such as *Milorganite* (a trade name) have been used effectively for home landscaping and gardens. They contain low levels of plant nutrients but have the advantage of an

Table 4.3

Composition of Steer Manure Samples Taken From Eight Phoenix Feedlots, 1963

		Nitrogen		Ash	
Sample no.	Total moisture to 60C (%)	Oven dry (%)	Wet (%)	Oven dry (%)	Wet (%)
1	36	3.64	2.34	19	13
2	35	2.65	1.72	41	26
3	58	2.89	1.22	22	9
4	54	2.62	1.18	26	12
5	31	2.32	1.60	55	38
6	48	2.61	1.37	31	16
7	30	2.51	1.75	40	28
8	28	3.18	2.28	30	22

	pH	Soluble salts in % of solids		Sodium	
Sample no.	Sat. paste (%)	Oven dry (%)	Wet (%)	Oven dry (ppm)	As % of soluble salts (%)
1	6.5	11	7	9,550	9
2	7.7	9	6	9,800	11
3	6.9	13	6	12,280	9
4	7.4	12	5	9,660	8
5	8.1	6	4	5,620	9
6	7.3	9	5	6,200	7
7	7.5	7	7	7,660	8
8	7.1	12	9	9,660	8

organic base which imparts a beneficial quality to soils needing improvement in structure and tilth.

Composts from plant residues also improve the physical properties of soils when applied in sufficient quantity. Every homeowner may make his own compost unit. Success in making compost in the desert Southwest depends on keeping the material moist and adding small amounts of chemical nitrogen fertilizer such as urea, ammonium nitrate, or ammonium phosphate. Occasional turning and mixing hastens the process. Degradation should proceed until the compost is crumbly and loses its obvious original identity as plant residue. Danger of perpetuating plant diseases is always associated with compost-making. Diseased residues are best discarded. Almost any plant residue decays into compost. Composts contain small percentages of nutrients, particularly nitrogen. Addition of nitrogen fertilizer along with compost provides an excellent combination for a fertilization and soil conditioning program.

Wood products such as sawdust and bark make good mulches. Mixed into soil, they improve the pore space and often the water-holding relationships in clays and heavy (fine-textured) loams. They supply very few nutrients. In fact, they may rob the soil of nitrogen as they degrade. Addition of nitrogen fertilizer in small amounts, as the materials decay, improves their quality as soil conditioners and prevents robbing plants of nitrogen. These residues should be brought up to about 1 or 1.2 percent nitrogen and kept until decomposition has proceeded for at least three or four months. For sawdust, this would equal between four to five pounds of ammonium sulfate per 100 pounds of dry sawdust.

Municipal composts have been used for growing plants in containers, in greenhouses, and in the field as well as around the home. They, too, contain low amounts of nutrients and, unless specified as being fortified with nitrogen, require the addition of nitrogen for best results. Municipal composts are derived principally from waste paper and have appeal from the aesthetic (less repulsive odor) as well as utilitarian standpoint.

Peat moss or *sphagnum moss* are tried and proven organic materials. Most growers mix them into soils to improve physical conditions and water relationships. They contain virtually no plant nutrients. Nitrogen supplementation is required.

Blood and fish meal from slaughterhouses, canneries, and fish markets contain protein, and, therefore, nitrogen. These products usually cost too much for large gardens but are effective as fertilizers for small planters and potted plants. Because of their relatively high levels of nitrogen, they are used sparingly.

Bone meal is an "old-timer." It supplies lime (calcium) and some phosphorus for acid soils. Soils of the desert Southwest, however, are already well-supplied with sufficient lime and, usually, with phosphorus.

Leather wastes from the hide industry have been suggested as organic fertilizers, but these decompose so slowly they contribute little to the nutrition of plants. They have a certain soil conditioning value, however.

Green manure derives its name from the green, succulent, and immature nature of the material of which it is composed. It is any plant residue which is incorporated into the soil while still in the growing stage. Most commonly used green manure plants are grasses such as ryegrass, wheat, or barley, and legumes such as alfalfa, Papago pea, clovers, and sesbania. The legumes are best-suited for green manure because of their very useful capacity in making nitrogen available, and therefore providing a favorable balance of nitrogen for the succeeding crops. In the desert Southwest, sesbania and Papago peas have been used with great effectiveness. Because fertilizing with green manure often is costly where water is limited, the practice under such condition has been abandoned in favor of commercial fertilization and turning under mature crop residues.

Mature crop residues. The renewal of organic matter in soils of the desert Southwest is largely dependent upon the utilization of mature plant residues. Mature plant residues of all sorts are valuable not only because they contain plant nutrients, but because their decomposition is slower than that of green residues, and their beneficial decaying extends over a longer period of time.

As its food supply is used up, the size of the soil microbial population diminishes, and nutrients contained in the dead microbial tissues are released for plant use. Nitrogen in the plant materials, largely in the form of proteins, is also utilized by microorganisms. As microbial activity slowly diminishes and decomposition proceeds to completion, the nitrogen of bacterial tissues (and this is considerable) is mineralized to ammonium and nitrate available for plant use. Burning of crop residues is definitely an unwise practice under normal conditions. Only in the case of pest or disease control should it be practiced, and then only as a last resort.

In burning off crop residues, soils not only are deprived of structure-building and moisture-conserving materials but are robbed of fertilizer elements. In terms of nitrogen, phosphorus, and potassium, crop residues are valuable fertilizer sources. Table 4.2 indicates the fertility values of certain residues. Crop residues are also a very valuable source of minor elements.

Turning under plant wastes often results in immobilization of nitrogen to such an extent the succeeding crop may show nitrogen-deficiency symptoms. This is due to the great demand by soil organisms for nitrogen during decomposition.

The need for wise utilization of residues should be again emphasized. It has been pointed out that these materials make an essential contribution to the productivity of soils during the process of decomposition. Only by decomposition can organic residue and organic matter function to a maximum extent for the benefit of plants.

Irrigated soils of the desert produce no less plant residue than most soils in the United States, including the prairie area, and most have as much residue returned to the land. The obvious difference between the condition in desert soils and those of the northern states is not in the *route* but in the overall *rate* of transformation of returned residues from whole plant constituents ultimately to carbon dioxide, water, mineral elements, and microbial substances, this being more rapid in desert soils owing to higher temperatures year round.

Management

Management of organic residues centers around how to prepare composted materials, method of incorporation or placement into the soil, and nitrogen-carbon relationships.

Compost Preparation

Composts may be made from almost any plant or animal residue. The homeowner can readily prepare any of three types in his own yard. These are: (1) compost prepared strictly with plant and/or animal residues, (2) compost prepared from vegetable or plant residues but fortified with commercial inorganic fertilizers and/or soil conditioners (often called artificial measures), and (3) compost prepared with animal and/or plant residues but layered with topsoil.

The procedure is much the same for the three types. *Type 1* is prepared as follows: Plant and/or animal residues are stacked in a pit dug into the earth, or piled on the surface. The pit has the advantage of preventing excess drying, which slows the rate of decomposition. In either event, plant residues such as straw, grass clippings, leaves, flower stalks, and yard prunings, household vegetable residues (potato peelings), and green weeds are placed in layers about three or four inches thick. Ammonium phosphate or superphosphate is sprinkled between each layer at a rate of about one pound per four water pails of material. Each layer is wetted thoroughly with a fine mist spray of water. *The composting material should be kept wet at all times.* Turning and mixing (every two to three weeks) hastens the process. Excessive turning, however, does not allow heat to generate sufficiently to kill diseases and insects. Fresh material may be added from time to time as convenient. The composting is complete when the material breaks and crumbles sufficiently to add easily to the soil. Completion of the compost process is more rapid in summer than in winter. Pet manures can be mixed into the compost pile if mixed well with the plant materials, and incorporated in such fashion that they create a minimum fly problem.

Horse manure compost does not require the addition of nitrogenous fertilizers. Its primary requirement is turning and occasionally sprinkling to keep the outer layers moist. Superphosphate and sometimes gypsum improve the quality of the compost. Superphosphate helps lower the loss of ammonia, and gypsum aids in granulation.

Poultry guano compost is prepared the same way as horse manure compost. More frequent turning and more attention to moisture content is necessary with poultry compost-making to prevent anaerobic odors and to keep the fly nuisance to a minimum.

Type 2 compost is more like animal manure, although it is made from plant material. It is well-fortified with commercial fertilizer. Woody materials such as ground bark and sawdust may be used, as well as other plant materials mentioned in Type 1. An old tried and proven formula for chemical mix to add is: ammonium sulfate (20% N), 45 parts by weight; superphosphate (20% P_2O_5), 15 parts by weight; and ground dolomitic limestone, 40 parts by weight. Gypsum is used in place of the limestone for desert conditions. This mix is sprinkled in the plant residue at a rate of one pound per water pail of dry matter. About half of this amount is used when the plant material is green and succulent.

Fresh *sawdust* and *bark* often contain growth-inhibiting substances. They should be composted for four to six weeks before use in gardens and plants in containers. Sawdust and bark cannot be expected to decompose perceptibly in even six weeks. In fact, about all that can be expected of them is to "cure-out" (i.e., eliminate the growth-inhibiting factors). Since these wood residues decompose so very slowly, they are not suitable compost material unless one wishes to wait several years for them to measurably break down.

Type 3 compost is prepared as Type 1, except the layers of plant residue are alternated with thin layers of topsoil and manure. The repeating sequence is layer of plant residue, layer of soil, and layer of manure. Commercial fertilizers may be added also to Type 3 compost. When manure is not available, kitchen waste may be substituted. Again the fly problem and possibly odors can make this type of composting a problem if watering and aeration are not controlled. Odors occur when the materials are too wet and oxygen (air) does not have an opportunity to penetrate the compost heap.

Composting is an excellent way to make use of polluting organic wastes, since almost any biodegradable material is suitable for processing into compost.

Placement of Organic Materials

The purpose of using organic residues decides the method of application. *Mulches* require no special techniques or equipment in their use. They are placed on top of the soil at a desirable depth over seedbeds or around plants. The possibility of mulches floating away during irrigation or heavy summer storms makes it necessary to provide means for holding them in place. Mulches function to (1) aid in conserving moisture, (2) modify soil temperature in the upper few inches where root activity concentrates, and (3) keep the soil surface open and porous, thus favoring water intake and infiltration and seedling emergence and establishment. Some growers feel that mulches encourage earthworm activity, which is beneficial to aeration and moisture relations.

Soil conditioning with organic materials is an old and trusted practice. Gardeners and farmers have long known that organic residues and wastes worked into the soil improve its tilth, aeration, and moisture conditions; the soil becomes loose and friable. The object is to incorporate the residue as deeply into the root feeding area as possible and mix with the soil. Home gardeners spade down compost, trash, manures, etc., as early as possible in spring, or even in the fall, to permit time for decomposition before planting. Farmers plow stubble and other residues into the soil well before the time of seeding.

In the desert Southwest such materials as peat moss, compost, and manures are mixed with the soil at or near the time of planting. Sandy soils with a tendency for droughtiness respond to fairly large quantities of organic materials. Planters support plant growth for long periods with mixes of equal parts of organics, clean sand, and good loam. Municipal and home composts are mixed in amounts equivalent to or greater than well-rotted manure. Sawdust can be used up to two-thirds by volume in a good loam. Nitrogen starvation may take place when a large proportion of the soil mix is organic material low in nitrogen. The addition of chemical nitrogen relieves this problem. Soil conditioning is treated in more detail in Chapter 6.

Fertilization with organic materials requires a knowledge of their chemical composition. Usually fertilization is done with materials of higher nitrogen content such as blood or fish meal. They are mixed into the seedbed before planting or topdressed around established plants. Large shrubs and trees are fertilized by drilling holes into the soil within the dripline of the plant and filling with the fertilizer.

The rate of decomposition is controlled by management factors — *aeration, moisture, temperature, nutrients* and *maturity of the residue.* Water being the most limiting factor in the desert Southwest makes management all the more important. Organic materials do not decay without moisture. With adequate moisture the rate of decay takes place rapidly throughout the year because of the favorably warm temperature and absence of winter refrigeration. Decomposition takes place most rapidly between 80 and 100 F. Bare soil surfaces in the desert exceed 100 F during the day in summer months. Decay gradually decreases below 80 F until it virtually stops at freezing. Nitrification, i.e., conversion of ammonium to nitrate or protein to nitrate, does not occur below soil temperatures of about 42 F.

Carbon-nitrogen relationships are critical in the growth of plants where large amounts of organic residue become mixed into the soil. Since microorganisms control nitrogen metabolism in the soil, plants get only the nitrogen that exceeds the microorganisms' needs for decomposition. Mature residues such as straw, sawdust, and leaves possess a wide carbon/nitrogen ratio: 80:1 or above. Decay of organic residues proceeds more rapidly with those materials that have a ratio of 30:1 or below than with those above 30:1. Management includes provisions for nitrogen supplementation. A good rule to follow is: apply nitrogen with all mature plant residues at the time they are mixed or spaded into the soil. Generally there are sufficient other nutrients such as phosphorus, potassium, and micronutrients, either in the residue or soil, to avoid these becoming limiting to plant growth during the process of decay.

As decay proceeds, nutrients immobilized by the microorganisms and those bound in the residue are slowly released to the soil in excess of the organisms' need. These become available for plant use.

Nitrogen applications often are associated with the conservation and maintenance of organic matter. This is due to the added increase in crop residue as a result of increased growth. Root as well as top growth is greatly increased by favorable nitrogen additions.

Maintenance

Desert soils are notoriously low in organic matter under virgin conditions, yet they are productive. Indeed some of the highest crop yields in the nation have been recorded from the soils of the desert Southwest. These soils often show an increase in organic matter when placed under irrigated agriculture. This is brought about by the vast amount of crop residues entering the soil as compared with those entering under virgin conditions. Soils of home gardens, lawns, golf and municipal turfs, and landscaped areas in general are often considerably higher in organic matter than the same soils are when under natural conditions in the absence of irrigation.

Certainly the improved organic matter content of home soils and turfs is a real factor in their permanent productivity. Favorable moisture conditions under irrigation, good aeration in desert soils, and warm temperatures favor rapid decay. This rapid rate of decay makes our attempts to increase organic matter content permanently, much above that of virgin conditions, economically impractical except in home gardens, planters, and certain turfs. It is the maintenance of this amount, however, that is so critical to continued favorable growing condition of the soil.

Desert soils need replenishment more often and at greater levels than humid soils if good tilth is to be maintained. Low organic matter content accentuates the need for good management and replenishment. Growers seldom apply less than ten tons of organic matter per acre and most often twenty tons per acre or about 0.5 to 1.0 pound per square foot.

5. What Nutrients Are Needed?

What is most surprising about plant nutrition is that out of 103 known elements in the universe, only 16 are considered to be required by plants. Moreover, those most abundantly used are light elements, like hydrogen, carbon, nitrogen, and oxygen with atomic numbers of 1, 6, 7, and 8, respectively. The remaining 12 elements account for but a small fraction of the total substances in plants. The heaviest of the required elements is molybdenum, which is needed in such minute quantities that it can be supplied mostly from the dust of the air. Over half of the known elements not needed have atomic weights greater than that of molybdenum. Perhaps what makes this so remarkable is that the soil contains most of the 87 elements not required, and the plant roots contact them throughout the life cycle.

Contrasted with the 16 known essential plant elements, listed in Table 5.1, are the larger number of 24 essential animal elements. Those essential for animals, but not for higher plants, include iodine, cobalt, chromium, selenium, and four elements used in "trace-trace" (very, very small) amounts — fluoride, silicon, tin, and vanadium. Dust in the air is sufficient to supply these. Perhaps some day these elements will be found to be necessary for plant growth. Fortunately for animals, plants take up elements not demonstrated as essential to plant growth. In this way the discrepancy of eight essential elements between the two biological kingdoms is resolved, and the nutrient requirements of strictly vegetarian animals are met. Silicon has an interesting role. Though it is needed for structure stability in plants, such as for stiffness of straw (many plants would lose their rigidity without silicon), it is not claimed to be essential. On the other hand, it is not essential for structure stability in animals, though it is essential for normal development of bones in chicks and possibly in other vertebrates.

Nitrogen, phosphorus, and potassium are absorbed in relatively large amounts (Table 5.1). They are deficient in many soils over the earth, particularly under intensive planting conditions or in sandy, heavily leached soils, and are therefore supplied through the use of fertilizers. All native soils of the desert Southwest require nitrogen; some need supplements of phosphorus; very few require potassium. When used for irrigated plantings, desert soils require an abundance of nitrogen to support maximum growth.

The need for N, P, or K in fertilizer is controlled by one or more of the following factors: (1) insufficient amounts present in the soil to meet the need of intensive plantings; (2) slow availability of the element; the soil is therefore unable to meet the need for the element by the plant during its maximum period of growth; and (3) inadequate balance of these three elements with each other or with other nutrients.

The kind and amount of plant nutrient needed in a soil may be predicted by sending a sample of the soil to the Agricultural County Agent's office or Farm Adviser to be tested. For the most part, plant nutrients are absorbed through *root hairs* from the soil solution (Fig. 5.1). Therefore the aqueous soil solution may be extracted, and the nutrient content determined chemically for soil fertility evaluation purposes. Complete correction of N-P-K defi-

Table 5.1

Plant Nutrients Required for Growth

Group 1	Those from air and water in large amounts:		
		Carbon (C)	
		Hydrogen (H)	
		Oxygen (O)	
Group 2	Those from the soil in relatively large amounts:		
		Nitrogen (N)	Calcium (Ca)
		Phosphorus (P)	Magnesium (Mg)
		Potassium (K)	Sulfur (S)
Group 3	Those from soil in relatively small amounts:		
		Iron (Fe)	Zinc (Zn)
		Boron (B)	Copper (Cu)
		Manganese (Mn)	Chlorine (Cl)
		Molybdenum (Mo)	Sodium (Na)

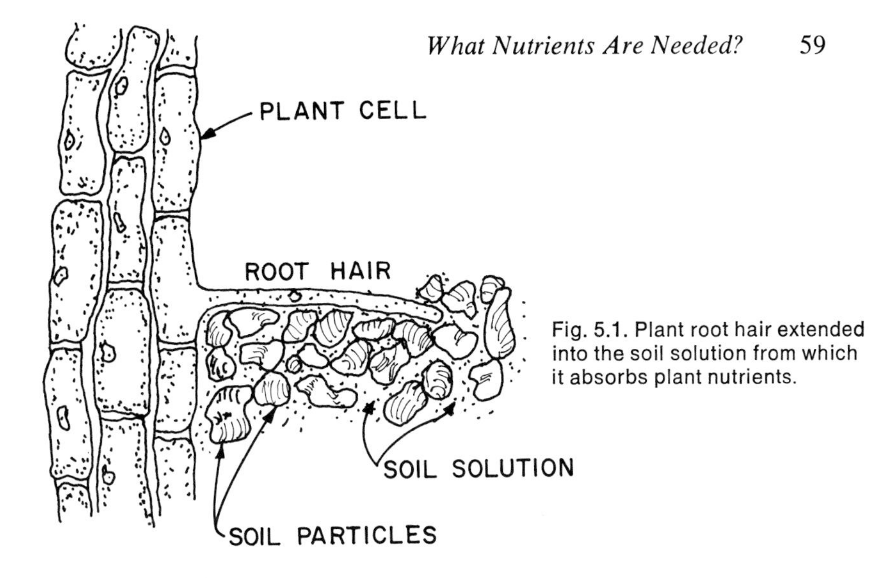

Fig. 5.1. Plant root hair extended into the soil solution from which it absorbs plant nutrients.

ciencies in plants can be made by wise use of commercial fertilizers and/or organic composts and manures.

Calcium, magnesium, and sulfur in Group 2 (Table 5.1) are also necessary to plants in relatively large amounts. These elements, however, occur in abundance in desert soils and rarely need to be added as fertilizer.

Nutrient elements in Group 3 (Table 5.1) are required by plants in such small amounts they are called *micronutrients* or *trace elements*. Plants do not grow normally without these elements, even though they are needed in very small quantities. Most micronutrients are found in minute quantities in plant tissues. All except sodium, boron, and chlorine react readily to form highly insoluble compounds in soils, and their availability to plants may be so low in certain soils as to constitute a deficiency. In such instances, available forms must be added to obtain suitable plant growth. Only small amounts (as little as a few ounces per acre) of micronutrients appear within the plant. To correct serious deficiencies, however, the amount of micronutrient fertilizer needed may be very much greater, because of the soil's tendency to render certain micronutrients unavailable to plants.

Sodium appears to be essential to only a few plant species. However, as research techniques improve, it is possible other plants will be found requiring sodium, and perhaps new elements will be added to the list of those essential for growth.

Nitrogen

The stimulating effect of nitrogen on plant growth is well-recognized. A vast amount of knowledge has been accumulated on nitrogen reactions, transformations, accumulations, and losses in soils. Since nitrogen fertilization and nutrition are so sensitive to variation in soil and water management, number and kind of plants, temperature, rainfall, soil texture, structure, aeration, and microbial activity and organic matter, no fixed quantity of nitrogen should be applied to all soils and all plants. However, by knowing the level of soil nitrogen and certain basic principles of nitrogen behavior and the requirements of the plants or crops to be grown, one can recommend a fairly effective nitrogen-management program for a specific plant and location.

Native Soil Nitrogen

The natural level of nitrogen in the soil is basic information that must be known before nitrogen recommendations can be made. Nitrogen in soils of the desert Southwest is lower than in those of other climates because organic matter, the carrier of nitrogen, is characteristically low. Additions of nitrogen fertilizers to native desert vegetation have been shown to encourage growth. Nitrogen is also the plant nutrient most universally needed in irrigated soils to produce high crop yields, good turfs, lawns, and productive and beautiful home gardens. Approximately 200,000 tons of nitrogen or nitrogen-carrying fertilizers were used annually in Arizona during 1965 to 1970. Because of its susceptibility to loss by leaching, denitrification to gases, and removal by crops and plant residues, nitrogen must be replenished regularly.

Native nitrogen in desert soils is primarily in the organic form as a part of the soil organic matter. Very little mineral nitrate (NO^{-}_{3}) is present at any one time in unfertilized soils. Small but continuous amounts are released by microbial conversions from insoluble organic matter sources, but plants absorb it rapidly for growth. During the last half of December and most of January and February very little nitrate-nitrogen is formed by microbial action.

Ammonium nitrogen (NH^{+}_{4}) is either not detected in our native desert soils or is present only in traces, because of its high rate of conversion to nitrate by the soil microflora, its ready absorption by plants, and high rate of volatilization into the atmosphere.

Transformations of Nitrogen

Nitrogen undergoes numerous changes in the soil. These changes are broadly categorized for convenience: immobilization, mineralization, nitrification, denitrification, fixation, and translocation. The nitrogen cycle is shown in Figure A.2, Appendix. Only by an understanding of these processes can nitrogen be integrated with the soil-water-plant systems in an effective and efficient program for better plant growth.

Immobilization. Nitrogen exists in the soil primarily in an immobilized state of organic combination as a component of plant, animal, and microbial residues and of microbial cells and their synthesis and degradation products. Organic nitrogen substances represent a pool from which available mineral forms (ammonia and nitrate) originate as a result of gradual microbial decomposition. The cycle becomes complete when plants absorb the mineral nitrogen and again incorporate it into their tissues as organic compounds, principally proteins. Immobilization of mineral nitrogen in the soil occurs as microorganisms utilize it for growth and reproduction during the decomposition of plant residues. When large amounts of residues, such as straw, leaves, and stalks, which have a carbon-to-nitrogen ratio greater than 30 to 1, are put into the soil, serious nitrogen deficiency symptoms may appear in young plants. This situation is remedied by adding nitrogen fertilizers either to the soil or residue.

Mineralization. During the decomposition of organic matter, nitrogen is liberated for plant uptake only when the supply is greater than that required for use by the microorganisms. Nitrogen will continue to be recycled by the organisms until the carbon content decreases to a level where the C/N ratio falls below about 30:1. This lowering of the C/N ratio takes place through the evolution of carbon, as carbon dioxide gas, to the atmosphere. Nitrogen, on the other hand, recycles until it is no longer needed for carbon degradation, at which time it is mineralized from proteins to the inorganic form in increasing amounts, primarily as ammonia, and then released to the soil where it is oxidized to available nitrate.

Nitrification. The oxidation of ammonia to nitrate in the soil is called nitrification. This process takes place readily when temperature and moisture conditions are favorable. Favorable moisture and temperature conditions correspond approximately with those for growing plants.

Oxidation of ammonia to nitrate proceeds in two major steps: (1) conversion of ammonia to nitrite by *Nitrosomonas* bacteria, and (2) conversion of nitrite to nitrate by *Nitrobacter* bacteria.

Nitrate formation takes place over a wide range of conditions, and is known to occur quite readily. Calcareous desert soils possess a very active nitrification microflora in all but the most alkaline soils, where sodium dominates calcium. Thus, nitrate is the dominant inorganic nitrogen form in soils of the desert Southwest. Ammonium from fertilizers added to soils also is readily converted to nitrate. With large applications, the toxic effects of excess ammonium on nitrite accumulation in desert soils is less than what would be expected in acid soils of humid regions.

Denitrification. Denitrification (nitrogen loss) is the process by which nitrates are reduced to elemental nitrogen gas (N_2), which escapes to the atmosphere where it is lost for plant uptake. Under conditions of limited oxygen supply, such as exist in waterlogged and compacted soils, and where large amounts of plant residues are incorporated, denitrification is pronounced. Overwatering can result in nitrogen loss if the soils are maintained wet and saturated too long. Because of the limited rainfall in desert climates and the unlikelihood of waterlogging, any great loss of nitrogen by this route would not be expected to be as great as under humid conditions. Loss of nitrogen from desert soils is primarily a result of ammonia volatilization.

Nitrogen fixation. Before plants can use elemental nitrogen gas of the air, it must be combined with other elements, such as hydrogen or oxygen. Such combining is termed "nitrogen fixation," and the resulting compounds are available to plant roots. It may occur through chemical reactions, as in the industrial production of ammonia and nitric acid, or biologically by the soil microorganisms. The conversion of atmospheric nitrogen to ammonia and finally to amino acids by soil microorganisms and legume plants through symbiosis in the root nodule is a good example of the latter. All desert soils contain nitrogen-fixing organisms, such as: (1) *Rhizobium*, a legume-nodule bacteria when in symbiosis (beneficial coexistence) with legume roots (nodules on roots, as seen in Figure 5.2, are the seat of N-transformation), (2) *Azotobacter*, free-living anaerobic bacteria, (3) *Clostridium pasteuranium*, a free-living aerobic bacteria, and (4) *Alga*, a free-living, aerobic, very small green plant (individual cell or plant can be seen by microscope).

Fig. 5.2. Effective nodulation on peas, *Pisum sativum* (courtesy the Nitragin Co., Milwaukee, Wisc.).

All, except possibly *Azotobacter*, are known to contribute substantially to the nitrogen economy of the soil. The contribution of *Azotobacter* is still somewhat obscure.

Translocation. Movement of nitrogen in the soil depends upon the chemical form in which nitrogen exists. For example, nitrogen as ammonia does not move readily in the soil because it is positively charged and is absorbed to the soil colloids (clays) sufficiently to prevent appreciable movement by the leaching action of water. Nitrate, on the other hand, is not absorbed to soil colloids as tightly and is highly mobile and moves readily in the soil water. Thus it may be lost to plants by leaching below the root zone or active root feeding area. Most plants can use both ammonium- and nitrate-nitrogen. Unfortunately, ammonium-nitrogen is rather rapidly converted to the more mobile nitrate form in desert soils. Excessive watering of soils to remove salts by leaching can result in considerable downward movement and loss of nitrogen fertility from the root feeding zone. Except when there is a need to flush the soil of salts that have accumulated in excess before or after the growing seasons, excess watering should be avoided. Nitrogen fertilizer is best applied *after* the pre-season leaching.

Nitrogen in Plants

Plant cells require nitrogen for their formation and functions. The nucleus, for example, is a nitrogen-containing structure of living plant cells. Nitrogen is also essential in the structures of chlorophyll and the formation of protein compounds.

Nitrogen occurs primarily in the organic form in plants, although variable quantities may be found in the inorganic form as ammonium, nitrite, or nitrate. The percentage of nitrogen in the different plant parts will vary with the age, type of tissue, kind of plant, and even the time of day.

Excluding the leaves, the aerial parts of most plants contain more nitrogen than the roots. Nitrogen content varies most in the leaves and least in the roots.

Plant response to nitrogen includes (1) encouragement of vegetative growth (stems and leaves), (2) assurance of a favorable rate of growth and plant development, (3) an increase in the intensity of green coloring, (4) an increase in the protein content of various plant parts, and (5) favorable seed production.

Adequate nitrogen. An adequate supply of available nitrogen is essential for maximum plant growth. Either an excess or deficiency will limit production.

Abundance of nitrogen. An abundance of available nitrogen in the soil produces a rank growth of foliage, stem, and leaves, and stimulates vegetative growth at the expense of flower, fruit, and often root development in certain plants. Unusual increase in leaf *area* is one of the most striking effects of an abundant supply of nitrogen. Nitrogen also is one of the most important factors in the *growth rate* of the leaves. The size of the plant is thus largely a measure of the rate of nitrogen metabolism.

With an abundance of nitrogen, the water content of tissues of various plants is increased. Such an increased succulence in the plant parts is considered to be caused by increased production of protoplasm which is highly hydrated, and a lower rate of transpiration in those plants receiving a high nitrogen supply compared with those having a limited supply. High nitrogen fertilization, therefore, makes plants more susceptible to freezing. It should be withheld late in the growing season, just as is water, to help "harden off" plants in preparation for cold weather.

Excess of nitrogen. An excess of available nitrogen salts in the soil will kill the plant. Plants wither, turn brown, and dry up. Excess of soluble nitrogen is particularly hazardous to seedlings when placed too close to the seed row by the sidedress method of application. The toxicity is due to excess of total salts, though, rather than to any specific toxicity of nitrogen itself. When plants are not killed, excess nitrogen limits the extension of roots and root development.

If plants can be grown to maturity, the excess foliage resulting from an abundant supply of nitrogen has no apparent effect on the quantity of fruit or seed. On the other hand, plants under a condition of excess nitrogen are delayed in maturing, and ripening occurs prematurely before sufficient food materials can be transferred from the vegetative parts to the seeds or grain. The seeds or grain thus produced are shriveled.

Deficiency of nitrogen. An insufficient supply of available nitrogen results in light green or yellowish leaves, dwarfed or stunted plants, limited branching of annuals, and small, poor-looking blooms and fruit. Lower leaves on the stem dry and drop earlier than usual. A decreased amount of protoplasm is formed, and a general reduction in leaves and stems occurs. Crop yield declines rapidly when nitrogen is deficient.

Forms Utilized

Nitrogen in the soil available to plants may be separated for convenience into two classes: inorganic and organic. The atmospheric sources available are: combined and elemental. The most common forms of soil nitrogen absorbed directly by the plant are ammonium (NH^{+}_{4}) and nitrate (NO^{-}_{3}). These ionic forms originate from inorganic salts (fertilizers) or as a product of organic matter decomposition.

Some plants grow equally well with either nitrates or ammonium salts as a source of nitrogen, and others, although assimilating ammonium salts in the absence of nitrate, seem to grow better when

Table 5.2

Approximate Nitrogen, Phosphorus, and Potash Content of Certain Harvested Crops

Crop	Estimated yield per acre	Nitrogen	Phosphorus		Potash	
		lb N	lb P	lb P_2O_5	lb K	lb K_2O
Barley — grain	2000 lb	36	6.99	16	9.13	11
— straw	1 ton	15	2.18	5	24.90	30
Corn — grain	5600 lb	90	15.28	35	20.75	25
— stover	3 tons	70	10.92	25	78.85	95
Sorghum — grain	3360 lb	50	10.92	25	12.45	15
— stover	3 tons	65	8.73	20	78.85	95
Wheat — grain	2400 lb	50	10.92	25	12.45	15
— straw	2 tons	30	3.49	8	41.50	50
Alfalfa — hay	6 tons	280*	26.20	60	232.40	280
Coastal Bermuda grass	8 tons	185	30.56	70	224.10	270
Beans — dry	1 ton	75	10.92	20	20.75	25
Oranges	5600 lb	85	13.10	30	116.20	140
Potatoes — tubers	400 lb	80	13.10	30	124.50	150
Tomatoes	15 tons	90	13.10	30	99.60	120
Cottonseed & lint	1500 lb	40	8.73	20	12.45	15
Sugar beets — roots	15 tons	60	8.73	20	41.50	50

*70 lb from soil; 210 lb from air

nitrates are applied. From the standpoint of the assimilation of these two forms, ammonium requires a lower expenditure of energy by the plant in its incorporation into protein than nitrate. There appears to be no apparent practical significance of this energy relationship. The effectiveness of one form as compared with the other appears to depend upon the *type* and *stage of growth* of the plant and the *acidity or alkalinity of the soil.*

Phosphorus

The amount of available phosphorus in desert soils usually is adequate for native vegetation in the southwestern desert. Plants native to either desert valley floors or dry uplands do not display phosphorus deficiency symptoms, nor do they respond well to phosphorus fertilizers in any appreciable degree. On the other hand, when soils are put into irrigation and the density of plants is increased, phosphorus is sometimes needed for some crops and certain landscape plants. Intensive use of the soils under irrigation conditions as compared with the native condition of the desert, as well as the difference in the kinds of plants involved, is responsible for the occasional need for supplemental phosphorus fertilizer.

Native Soil Phosphorus

Native soil phosphates are dominated by the calcium phosphates. Compared with soils of humid regions, only a small proportion of the soil phosphates are combined into iron and aluminum complexes or with kaolin clay. Since these forms are highly unavailable to plants as compared with the calcium phosphates of desert soils, phosphorus is not as limiting a plant nutrient in the desert Southwest as it is in humid climates.

Organic phosphorus also is found in desert soils in amounts ranging from about 10 to 20 percent of the total in the main root feeding area (Fig.5.3). Most of the organic phosphorus is in the upper part of the soil profile associated with the soil organic matter. See a diagram of the phosphorus cycle, Fig. A.4, Appendix, for a complete phosphorus budget.

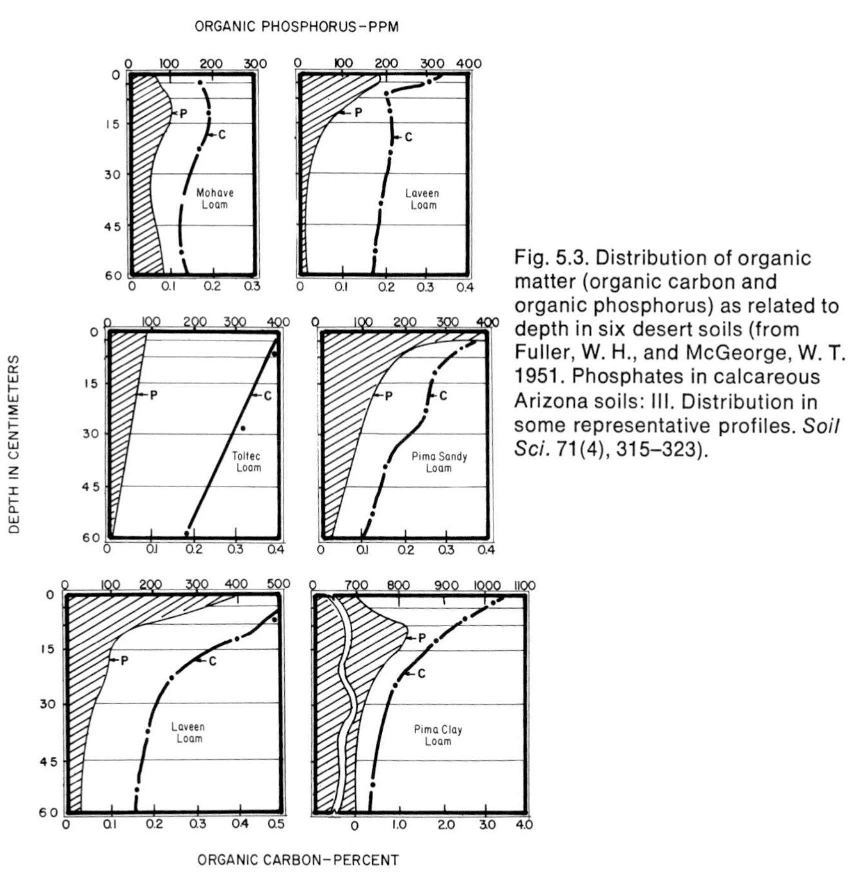

Fig. 5.3. Distribution of organic matter (organic carbon and organic phosphorus) as related to depth in six desert soils (from Fuller, W. H., and McGeorge, W. T. 1951. Phosphates in calcareous Arizona soils: III. Distribution in some representative profiles. *Soil Sci.* 71(4), 315–323).

Reactions of Applied Phosphates

Phosphorus reactions are unlike nitrogen reactions in soils. Phosphorus moves but little and thus is not lost from root zones by leaching water. Research has demonstrated that more than 90 percent of the applied phosphorus fertilizer moves less than one inch from its placement in soil. Liquid fertilizer phosphorus added to the surface was confined to the upper four inches, even in a porous sandy soil.

The slow movement of applied soluble phosphorus in desert soils makes it necessary to place the fertilizer down into the root feeding zone if it is to be of maximum use to the plant. Surface applications of phosphorus are of little value until they are plowed or spaded into the root zone.

Small amounts of phosphorus in the organic forms of microbial and plant debris have been shown to move downward, but the process is slow. Roots, decaying and dying in the soil, contribute to the distribution of phosphorus throughout the soil profile (as illustrated in Fig. 5.3).

Another reason for the occasional requirement of home garden plants and field crops for phosphate fertilizer is the removal of plant residues. Under natural conditions, plant residues remain in the soil where they are produced and therefore become a part of a phosphorus recycling process, concentrating phosphorus in the surface layers.

A large portion of the phosphorus in fertilizers added to desert soils reacts with the soil and is unavailable for immediate plant use, but becomes available to plants over fairly long periods. Only a small part is removed with plant residues each year; so, in general, phosphorus fertilization is not needed as frequently as nitrogen fertilization. Home gardens may not need phosphorus additions more often than once in two or three years.

Phosphorus in Plants

Phosphorus is primarily involved in the conservation and transfer of energy in living cells. Thus, all cells of the plant are dependent on phosphorus, and its distribution throughout the plant is governed by need. Phosphorus in plants is mobile and does not remain fixed like calcium, iron, and certain other nutrients.

Phosphorus is redistributed within plants when the available soil phosphorus becomes limiting. Phosphorus is withdrawn from the older, less active cells and transferred to the younger, more active cells. Later, the phosphorus is withdrawn from the leaves and transported into the fruit.

Deficiency symptoms of phosphorus are not specific. If phosphorus is needed, however, the following conditions may appear: (1) Root and shoot growth is greatly reduced. (2) Shoot growth is upright and spindly. (3) Lateral shoots are limited and lateral buds may die or remain dormant. (4) Premature defoliation occurs, beginning with older leaves. (5) Blossoming is reduced. (6) Leaf and blossom buds delay opening in the spring. (7) Foliage may become purple or spotted with bluish-green blotches having purple or brown centers. Leaf margins may turn brown.

Different plant species have different abilities to absorb phosphorus from soil or fertilizer. For example, cotton rarely responds to phosphate applications, in terms of lint yield, whereas legumes (peas, beans, lupine) often require some additional phosphorus during growth. Pyracantha and oleander respond infrequently, but vegetable crops respond frequently. Citrus trees may respond when young, but mature trees rarely show phosphorus response.

Examples of differences among plants with respect to phosphorus utilization are numerous. This leads to the postulation that chemical laboratory tests for available soil phosphorus necessarily must be related to the plant to be grown before recommendations can have practical application.

The need for available soil phosphorus in the early stages of plant growth is relatively great. The root system is small during the early stages of growth; thus the feeding area is limited and a greater concentration of phosphorus per root volume is needed. Young plants also demand more phosphorus than mature plants. It has been stated that plants usually have obtained 50 percent of their phosphorus needs when only 20 percent of their entire growth has taken place.

Fast growing plants maturing during the warm summer and fall months often have been found to respond more to phosphate applications than the same plants growing more slowly during the cooler months.

Forms Utilized

Plants have a capacity to absorb and keep great quantities of phosphorus in their tissues, in excess of their needs. The cells contain both inorganic and organic forms. Soils of the desert Southwest supply soluble calcium phosphates to plants.

Certain simple forms of organically bound phosphorus appear to be utilized by plants. Most of the evidence, however, points to the unavailability of organically bound phosphorus found in soils, and to the dependence on biological mineralization (or decay) of the organic phosphorus compounds to soluble inorganic forms before phosphorus is absorbed by the plant.

Potassium

Potassium is well-supplied in soils of the desert Southwest, and fertilizer additions are required only occasionally in sandy soils. The relative abundance of available potassium in desert soils is one of their most outstanding and distinguishing characteristics. Most residential and landscape plantings, however, have not been subjected to potassium deficiency tests. There is reason to believe some home plants which have been brought into desert conditions may have need for more available potassium.

Reactions in Soils

Studies comparing the behavior of potassium between arid and humid soils reveal the following:

1. Replaceable potassium (held on clay particles) is more available in alkaline-calcareous soils than in neutral- or acid-noncalcareous soils (see Fig. 4.3 for illustration of mechanism of replacement reaction).
2. Water solubility of potassium in calcareous soils is lower than in noncalcareous soils, except for those containing a large excess of alkalinity.
3. Solubility of potassium is greatly increased by carbonic acid in limy soils. Carbon dioxide (CO_2) in air + water (H_2O) yields → carbonic acid (H_2CO_3). In nonlimy soils, carbonic acid exerts much less effect on potassium and, in many cases, causes no increase in solubility at all. This latter observation possibly explains the greater potassium availability in limy desert soils.

4. Exchangeable potassium is more easily replaced from calcareous than humid climate noncalcareous soils. (An exchangeable element is one which is held to the clay or fine soil colloids by chemical bonds, but may be released to the soil solution by exchanging places with another element in the soil solution.) Plants can take up exchangeable elements, i.e., those elements held or absorbed on the soil particles and removed by exchanging with another element of similar charge.
5. Non-exchangeable potassium is higher in arid than non-arid soils.

Certain saline and alkaline soils have been found to have excessively large quantities of water-soluble and exchangeable potassium. Some of these soils require reclamation treatment, in which the excess potassium is leached below the root zone. The physical condition of the soil is adversely affected, and water penetration rate is slow in potassium soils, somewhat similar to the way soils containing sodium salts are affected. Fortunately, potassium soils are not found extensively in the irrigated areas of the desert Southwest.

Potassium in Plants

Potassium is one of the many elements considered essential for higher plant life. Although its specific physiological role is not clear, potassium has certain effects on plants. It is believed to play an important part in the following plant processes:

1. Carbon dioxide utilization: thereby association with the synthesis and translocation of carbohydrates, such as simple sugars and starches.
2. Cell wall formation: thereby aiding in rigidity of stems and turgor of cells. However, if nitrate reserves are excessive, plant cell walls may be thin and plants may have weak stems despite high levels of potassium.
3. Absorption of phosphate and nitrate under certain conditions.
4. Resistance to disease. This may be an indirect result of increased vigor, thereby improving the ability of the plant to resist attack by pathogenic organisms.

Potassium moves readily throughout the plant. It is prominently present in the actively growing tissues.

Plants deficient in potassium exhibit distinct, well-characterized symptoms: (1) Plants grow slowly. (2) Margin of leaves show "browning" or "scorching," starting first on older leaves. (3) Stalks are weak. (4) Seed and fruit are small and shriveled. (5) Clovers and alfalfa show characteristic white spots near leaf margins.

Plants differ in their overall requirement for potassium, as well as in their ability to absorb it from the soil. The potassium requirement may be more critical at one physiological growth stage than another. For example, potatoes may show no symptoms of potassium deficiency during the early stages of growth, but during later stages deficiency may become critical. This critical stage, however, is not necessarily the same for all plants.

Calcium, Magnesium, and Sulfur

Calcium, magnesium, and sulfur are well-supplied in soils of the desert Southwest. In fact, all three elements appear in excess of usual plant needs. Fortunately, these excesses are not directly toxic, as such, to plants. Their presence in excess, however, has been shown to influence the absorption and uptake of certain other nutritional elements from soils. Lime-induced chlorosis in plants, for example, results from excess calcium salts (principally calcium carbonate), which depresses the availability of soil iron to plants. *Chlorosis* is a yellowing of leaves caused by lack of green chlorophyll formation. Iron-deficiency symptoms may appear in high lime/caliche knolls, outcrops, and thin layers of topsoil over lime soil.

Rarely has it been found that high amounts of magnesium in desert soils have an adverse effect on plant growth. Usually, high magnesium soils adversely affect plant growth only indirectly by altering the physical condition unfavorably. High magnesium soils are claimed to possess some of the undesirable characteristics of high sodium soils, i.e., dispersed soil structure, compaction, poor aeration, slow water penetration, bicarbonate toxicity, and salt accumulation.

Sulfate (SO_4) is plentiful in desert soils; there is no need to add it. Sulfate must not be implicated in the beneficial effects of elemental sulfur (S) through acidification. Sulfur oxidizes to sulfuric acid. The acid improves the physical condition of soils and increases the availability of certain plant nutrients (see the sulfur cycle, Fig. A.5, Appendix).

Iron

Iron is the micronutrient most frequently observed to be deficient in calcareous desert soils. Although most desert soils contain large amounts of iron compounds, the compounds are quite insoluble and therefore are not readily available to plants. The symptoms of iron chlorosis are usually yellowing of the leaf area between the veins. In Figure 5.4, the peach tree has very yellow leaves in contrast to the pecan tree in the rear. Iron sulfate spray temporarily cured the iron deficiency, and the leaves turned green. Lime-induced chlorosis may occur as a result of deficiency of available iron where lime (caliche) is high. Soils of the desert Southwest having high alkalinity also may have a very low amount of iron available for plant use.

It is not uncommon to observe iron chlorosis in home gardens, lawns, and shrubs as a result of overwatering causing chemical changes in iron and closely associated compounds. Soils kept too wet by excessive irrigation become waterlogged and poorly aerated.

Fig. 5.4. Peach trees on high lime soil show yellow leaves due to iron deficiency. The center tree is a pecan which does not show iron deficiency chlorosis. Spraying with iron sulfate or chelated iron corrects the deficiency temporarily.

Plants on these and other soils where drainage is not adequate become chlorotic due to lack of iron availability, or ability of roots to absorb the iron. If the deficiency persists, the leaves of plants begin to turn brown and drop off, twigs die back, and eventually the plant dies. Trees and shrubs placed in holes above caliche with no provision for drainage from the bottom may turn yellow and eventually die from a lack of ability to absorb iron. Iron also may not be absorbed as result of antagonism by other accumulated salts.

Applications of large quantities of phosphate fertilizer over a period of several years have been known to lower the availability of certain micronutrients, particularly iron. Lawns are often observed to turn yellow in midsummer where ammonium phosphate has been added too frequently or in excessive amounts. This can be corrected by changing to another form of nitrogen fertilizer that does not contain phosphate.

The question often is asked, "Why is iron deficient for plants in desert soils when these same soils may contain as much as 8 to 15 percent calculated iron oxide (Fe_2O_3) and iron is the third most abundant mineral element found in the earth's crust?" The answer is that under certain conditions iron forms compounds that dissolve so slowly that plants cannot get enough iron to satisfy their requirements. Mohave sandy loam, for example, has been shown to contain 11 percent iron compounds calculated as Fe_2O_3, yet certain plants have been known to become chlorotic when grown in this soil. Much of the iron in desert soils is in the form of insoluble iron oxides and iron hydroxides or is immobilized in clay minerals.

Plants vary in their ability to obtain sufficient iron for their needs. Plants whose normal habitat is acid rather than alkaline soils are most susceptible to iron chlorosis. The usual range of iron concentration in plants is from 4 to 16 ppm.

Iron deficiencies have been corrected temporarily by use of iron sulfate. Either ferric or ferrous sulfate can be used, since both are about equally soluble in water. Foliar (above-ground parts) sprays are more effective than soil applications, although they too have transient characteristics. Iron sulfate is not usually recommended as a soil treatment for chlorosis in desert soils, because it quickly forms iron hydroxides and oxides unavailable to plants. Soluble iron from added sulfates also combines readily with many of the soil constituents to form other insoluble compounds such as phosphates, silicates, and carbonates.

Iron chelates have been developed that can be used effectively as soil amendments to correct iron deficiencies. The treatment lasts somewhat longer than iron sulfate. Chelates are organic compounds that protect the iron from becoming insoluble and unavailable to plants in soils and from precipitating with other elements. Plants can absorb the whole chelate compound and incorporate it into their tissues before iron becomes insolubly combined in the soil. Not all iron chelates, however, remain suitably soluble in alkaline calcareous desert soils. Only the chelates recommended for calcareous soils should be employed. The directions on the packaged compound give this information.

Zinc

Rarely is zinc found to be deficient in plants grown in soils of the desert Southwest, although zinc responses have been identified for pecans and certain fruit trees and responses occasionally have been reported for citrus. Home gardens, residential and urban plantings, and most vegetable crops have not appeared to need zinc, but no thorough deficiency study has been made of garden plants and imported ornamental plants and shrubs.

The same conditions that cause the low availability of iron also cause low availability of zinc, manganese, and copper. Briefly, these conditions are listed as: (1) presence of lime (calcium carbonate or caliche), particularly the calcite form, (2) presence of dolomite (magnesium-calcium carbonate), which also strongly absorbs zinc, (3) soil pH values in the neutral to alkaline range, (4) low oxygen level in soils due to excess water, and (5) excess application of phosphate fertilizer.

Fortunately zinc deficiency is not a serious problem. However, with the advent of more intensive use of the land, applications of certain fertilizers, and soils becoming older in terms of years under cultivation and crop removal, it is expected that micronutrients will become less rather than more available in desert soils. Zinc deficiency is likely to show up along with iron deficiency.

Manganese and Copper

Neither manganese nor copper have been found to be deficient for plants in arid soils. If deficiencies exist, the symptoms are so slight that they remain hidden. Manganese and copper are expected to be deficient under about the same conditions as iron and zinc.

Chlorine, Boron, and Molybdenum

Soils of the desert Southwest, as well as desert soils generally, are more apt to contain excesses of chlorine, boron, and molybdenum than deficiencies. There is no need, therefore, to add these elements. Excess quantities of chlorine and boron often occur in alkaline soils or as a result of poor irrigation management. Leaching with good quality water is necessary to remove excess chlorine and boron.

Nutrient Balance

A balance of micronutrients in the soil is just as important as a balance of macronutrients. Since desert soils often contain accumulations of nutrients far in excess of the plant's requirement, it is desirable to point out certain possible imbalances that can occur: (1) Iron deficiency may be caused by an excess of zinc, manganese, and copper. (2) Excess copper or sulfate may adversely affect the uptake of molybdenum. (3) Excess phosphate may cause a deficiency of zinc, iron, and copper. (4) Heavy nitrogen fertilization may cause copper or phosphorus deficiency. (5) Manganese uptake may be limited by high sodium and potassium content. (6) Boron uptake can be reduced by excess lime. (7) Manganese uptake may be diminished by iron, copper, and zinc excesses.

Because of these and other antagonistic effects, putting nutrient elements into the soil without knowledge of the soil's physical, biological, and chemical makeup can be hazardous. Indiscriminate use of micronutrients can worsen the soil productivity and reduce the beauty of home gardens and landscaping.

6. How Fertilizers Are Used

Fertilizer recommendations for humid climates are not suitable for arid climates. Both the *kind of fertilizer* and its *management* are different. Reasons for this are:

1. The pH reaction values of desert soils are neutral to alkaline, rather than acid.
2. Many nutritive elements such as magnesium, calcium, sulfur, chlorine, and, for the most part, potassium, are available to plants in adequate quantities.
3. Salts and certain toxic elements often are present in excess as a natural circumstance where rainfall is light and leaching is minimal.
4. Although soils of the desert Southwest are highly productive, there is an extra demand put upon them for soil nutrients, since in such a favorable climate they support plant growth throughout the year.
5. Water is the single most limiting factor for maximum plant growth, making irrigation necessary. Irrigation of soils complicates fertilizer management practices.

Fertilizer Grades and Ratios

A fertilizer may contain only a *single* nutrient element, or it may combine *several.* Ammonium nitrate, for example, contains only nitrogen as a soil nutrient. It is used when nitrogen alone is needed. Ammonium phosphate contains both nitrogen and phosphorus and is called a *mixed* fertilizer. A *complete* fertilizer contains all three major nutrients — nitrogen, phosphorus, and potassium.

Table 6.1

Some Commercial Fertilizers and Their Average Composition and Ratios of Nitrogen, Phosphorus, and Potassium

Name	Nitrogen	Phosphorus		Potassium	
	N (%)	P (%)	P_2O_5 (%)	K (%)	K_2O (%)
Ammonium sulfate	20.5	0		0	
Ammonium nitrate	33.5	0		0	
Calcium nitrate	15.5	0		0	
Calcium cyanamide	22.0	0		0	
Anhydrous ammonia	82.2	0		0	
Ammonia solutions	20.0	0		0	
Liquid (urea+ammonium nitrate)	32.0	0		0	
Ammonium phosphate	11.0	21	48	0	
Diammonium phosphate	21.0	9	53	0	
Ammonium phosphate sulfate	16.5	24	20	0	
Phosphoric acid	0	23–24	52–54	0	
Superphosphate, normal	0	8–9	18–20	0	
Superphosphate, treble	0	18–21	40–48	0	
Nitric phosphates	Variable	0	Variable	0	
Potassium nitrate	13.4	0		37	44
Potassium sulfate	0	24	55	41	49
Potassium metaphosphate	0	0		29	35

Mixed Commercial Fertilizers

Ratios* $N–P_2O_5–K_2O$	Percentages $N–P_2O_5–K_2O$	N (%)	P (%)	P_2O_5 (%)	K (%)	K_2O (%)
1–1.25– 0	16–20– 0	16	8.8	20		0
1–4.1 – 0	11–48– 0	11	21.0	48		0
1–2.0 –0.5	10–20– 5	10	8.8	20	4	5
1–2.6 – 0	18–46– 0	18	20.0	46		0
1–2.0 – 1	10–20–10	10	8.8	20	8	10
1–3.0 – 0	16–48– 0	16	21.0	48		0
1–2.5 – 0	21–53– 0	21	23.3	53		0
1–0.5 –0.25	15– 8– 4	15	3.5	8	3	4
1–1 – 0	24–24– 0	24	10.5	24		0

*Comparing N, P_2O_5, K_2O, using N as 1

The contents of fertilizers must by law be displayed prominently on the bag or container in terms of *grade*. A fertilizer grade (Table 6.1) refers to minimum guarantee of the plant nutrient content in percentages of:

Total nitrogen (N%)
Available phosphorus (P_2O_5%)
Water-soluble potassium (K_2O%)

For example, a fertilizer labeled 5-10-10, contains 5 percent nitrogen calculated as N, 10 percent P_2O_5, and 10 percent K_2O. Ammonium nitrate appears as 33.5-0-0, ammonium phosphate as 11-48-0 or 16-20-0, and potassium nitrate as 13-0-44. The elements always appear in the labeled grade or ratio in the same order: nitrogen *first*, phosphate *second*, and potash *third*. The fertilizer *ratio* is sometimes used to indicate the relation of one element to another. The ratio is found by calculating the relative percentages of the elements in whole numbers as near to unity as possible. For example, a 5-10-10 fertilizer has a 1-2-2 ratio of nitrogen as N, phosphorus as P_2O_5, and potassium as K_2O.

A newer concept is to report the elemental, N-P-K, content of a fertilizer and not N - P_2O_5 - K_2O, yet most of the agricultural chemicals industries are reluctant to make the change because of the historical significance of the older designation. The oxygen designation (i.e., P_2O_5 and K_2O gives the buyer the impression the fertilizer contains more phosphorus and potash than it actually does.

Commercial fertilizers are defined as substances and mixtures of substances containing 5 percent or more of nitrogen, or available phosphoric acid or soluble potash, singly, collectively, or in combination; except manures, hays, straws, peats, and leaf molds.

Nitrogen

Nitrogen is the single nutrient element which most often limits plant growth in soils of the desert Southwest. It must be supplied regularly to all soils when taken out of their virgin state and planted. The amount and kind of fertilizer needed will depend upon the supply in the soil, its availability, and the crop or plant to be grown. Although reserve supplies of nitrogen may be low, legumes (alfalfa, garden peas, beans, sweetpeas, lupine) that use atmospheric nitrogen through root nodule bacteria do not respond to nitrogen fertilization, whereas nonlegumes do.

Kinds of Nitrogen Fertilizers

The most popular nitrogen fertilizers, along with other fertilizers used in the Southwest, are listed for convenience in Table 6.1.

Ammonia is one of the cheapest and most abundant of the fertilizer nitrogen sources. Several forms are sold, such as: (1) ammonium sulfate (21% N), a granular form used universally (lawns, vegetable gardens, home and public landscaping, and agriculturally), (2) ammonium phosphate (16% N), a granular form used universally, and (3) ammonium nitrate (33.5% N), a granular form used universally.

These solid forms are more convenient and safer for use in home gardens and home and municipal landscaping, as well as in sidedressing small plants, than the commercial liquid or gaseous forms. In general, the solid nitrogen fertilizers are equally effective on an equal nitrogen content basis. For example, almost twice as much ammonium sulfate or phosphate must be added to supply the same amount of nitrogen as ammonium nitrate. The major distinction is the cost per weight of nitrogen. This may be calculated by dividing the percentage of nitrogen into the cost per 100 pounds of the fertilizer, as:

$$\frac{\text{Cost per 100 lb of fertilizer}}{\text{Percentage N in fertilizer}} = X$$

The smaller the factor X, the more economical the fertilizer nitrogen.

Where two nutrient elements, nitrogen and phosphorus, for example, are present, as in ammonium phosphate (16% N and 20% P_2O_5), the phosphate must be considered also in the cost. However, if phosphorus is not required for better plant growth, it does not enhance the overall utility or value of the fertilizer.

Urea, $CO(NH_2)_2$, an organic nitrogen source containing 45 percent nitrogen, reacts similarly to ammonium and is transformed either biologically or chemically to ammonia in the soil. A good proportion of fertilizer nitrogen used agriculturally is in the form of urea, partly because of its high nitrogen content and partly because of its low cost-per-unit nitrogen when used in preparing mixed fertilizers.

Ureaformaldehyde compounds have been developed because of their slow release of nitrogen and retention of same in soil as compared with other forms of fertilizer. They are mixtures of urea and formaldehyde which have been reacted under controlled temperature and pressure to form new, slightly soluble compounds. The nitrogen is supplied gradually over the growing season rather than all at once. The cost on an elemental nitrogen basis is higher than that of the highly water-soluble urea. However, savings from loss by washing below the reach of roots may compensate, in part at least, for the higher cost in soils subject to extensive leaching.

Nitrate fertilizers are manufactured primarily through the oxidation of ammonia. Nitric acid is reacted with various substances to produce nitrate fertilizers. Several forms appearing on the market are: (1) ammonium nitrate: a number of liquid and solid preparations are well-suited for desert soils (Table 6.1), (2) calcium nitrate: suitable for desert soils, (3) potassium nitrate: used primarily where potassium is needed (it is seldom needed in the desert Southwest), and (4) sodium nitrate: not suitable for arid and semiarid soils because of the detrimental residual effect of sodium on soil structure.

Nitrates are highly soluble in water and move readily with water in the soil. Thus they may be lost to plants when leaching extends below the effective root zone. They therefore must be replaced several times throughout the growing season as supplemental applications.

Natural organics have been the backbone fertilizers of home gardens for centuries. Since the late 1930s they have had to compete with manufactured commercial fertilizers. The demand for fertilizer is so great that the supply of natural organics can satisfy only a small fraction of this demand. If it were not for commercial fertilizers, prices for food and fiber would be unreasonably high and famine would be rampant over much more of the world. Gardens would be struggling to survive. All-natural fertilizers would be required for food production, leaving little for home gardens and landscaping.

Organic nitrogen sources are used extensively around the home and in public landscaping, since homeowners can better afford to pay the premium price than can the agricultural user. Natural organic fertilizer materials such as animal manure, guano, compost, and sewage residue generally are low in nitrogen, that is, 5 percent or less. They are costly per unit of nutrient element as compared with commercial fertilizers.

Natural organics have several advantages, however, which appeal to home gardeners and landscape architects. Some of these advantages are:

1. They are safer to use when applied in excess, since they are less apt to burn or kill plants than high-analysis commercial fertilizers.
2. Nitrogen is released more slowly and over a longer time during the growing season.
3. Many of them have soil conditioning properties and improve the physical as well as the nutritional condition of the soil.
4. Many of them supply micronutrients or some other nutrient element, the deficiency symptom of which may be hidden and thus go unnoticed by the gardener.

Application

The *timing* of application of nitrogen fertilizer to the soil is important to make the most effective use of the fertilizer and prevent its loss by leaching. Factors to consider in effective timing are: (1) kind of nitrogen: ammonium, nitrate, or organic form, (2) kind of soil: sand, silt, clay, or organic, (3) soil temperature, (4) kind of plant and the season in which the plant makes its maximum growth, and (5) stage of growth: young plants need more nitrogen.

An example of the influence of the kind of nitrogen on time of application is in the use of ammonium and nitrate forms. In preplant applications, where loss by leaching could occur, ammonium fertilizers are preferred over nitrate forms. Nitrate forms are most effective in the early or midgrowing period when rapid growth is desired. Two or more light applications also are preferable to a single heavy application where soils are very porous or sandy, and leaching is prevalent.

Regardless of the kind of manure used (poultry, rabbit, horse, cow, steer, or household pet) or its state of decay, *it should be applied early*. Manure can burn plants but only if it is fresh or high in salt. Composted and well-rotted manures do not harm most growing plants if they touch them. Best results may be obtained by broadcasting the manure on the soil surface one-half to two months in warm weather and three to four months in cold weather before planting. Horse manure is low in nitrogen and high in fiber. Composting it until heating stops and the strawy matter is well-decomposed greatly improves its quality. Horse manure definitely should be mixed well with the soil before planting.

The same factors that influence timing of application also

affect *placement.* Mineral nitrogen fertilizers, if placed too close to the plant, can burn its roots. "Starter" fertilizers are most effective if placed in a *band* to the side and below the seed. After the plant is up, it is well to either add nitrogen in the *irrigation water* in a dilute concentration or irrigate it into the soil if applied *broadcast* to the soil surface. *Direct contact of mineral nitrogen fertilizers with the seed or plant should be avoided.* The best procedure involves working "*surface-applied*" fertilizer into the soil with some mechanical implement before irrigating.

Manures are most effective when mixed into the top six to ten inches of the soil, then watered to allow decay to proceed. Only by decaying can the organic-bound nitrogen be released for plant use. Well-rotted manures may be applied to new plantings of trees and shrubs by mixing them with the soil at the bottom of the hole and covering with soil. Rose roots must be protected from any contact whatsoever with manure of any kind. Roses and similar sensitive plants, however, will tolerate manure in the soil adjacent to the roots or as a surface mulch.

Best results may be obtained from organics low in nitrogen, such as manures, composts, mosses, bark residues, sawdust, or municipal sludges, by adding small amounts of high-nitrogen commercial fertilizer to them before placement.

Starter solutions are liquid fertilizers that contain combinations of nitrogen, phosphorus, and potassium or a complete plant food mix. They provide desirable growth stimulation for young plants and transplants. Starter solutions are prepared easily by dissolving soluble fertilizers in water at such a low concentration they will not harm plants but will supply an abundance of readily available plant food. They originated as a stimulant for rapid growth of transplantings (such as bedding plants) while the limited root systems become established. Starting solutions may be used as an aid in transplanting desert vegetation — cactus, yucca, creosote bush, pentstemon, etc. They also give struggling young plants an immediate boost when nutritional scarcity is discovered. Some nurseries carry organic starting solutions containing vitamin B12, as well as solutions of a complete N-P-K plant food. The primary benefit of these solutions is in their nitrogen, phosphorus, and potassium content. Home gardeners can make effective starter solutions cheaply and easily. Solutions containing nitrogen are preferred. Some examples are:

1. *N-P-K starter*: Dissolve one oz. (28.35 grams) of a 10-10-10 or 6-10-4 commercial fertilizer in one gallon of water.
2. *N-P starter (Ammo-phos)*: Dissolve one-half ounce (14.18 grams) of ammonium phosphate (16-20-0) in one gallon of water.
3. *N starter (nitrogen)*: Dissolve one ounce (28.35 grams) of ammonium sulfate (20-0-0) in one gallon of water.
4. *Organic starter*: Place one-half pint (one cup, 8 oz. or 113 grams, air dry) of well-rotted manure (free of excessive salts) in one gallon of water. Mix and allow to stand overnight before using.

Accurate measurement and weighing of the ingredients being used to prepare starter solutions is mandatory. Inaccuracy in the proportions of fertilizer to water leading to stronger solutions than above may damage or kill plants. These solutions are designed to be poured in the soil around the plant and *not on the foliage*. Applications range from one-half to one pint per plant.

Selecting a Nitrogen Fertilizer

A fertilizer should be selected to fulfill a particular situation. Though nutritional differences for the most part may often be slight, cost per quantity of nitrogen varies considerably. Organic fertilizers, for example, are more expensive per unit weight of nitrogen than inorganic or strictly chemical ones.

Differences among various sources of nitrogen fertilizer may be indicated according to: (1) degree of loss by leaching from the soil, (2) mobility or immobility of source, (3) effect of soil reaction (pH value) on availability, (4) residual effect of accompanying element or material not related to soil reaction, (5) carbon-nitrogen ratio of natural organics, (6) content of micronutrient, and (7) form of nitrogen required by the specific plant species.

Evaluation of Soil Nitrogen

The correct amount of nitrogen fertilizer to add to a soil is not easy to determine. Most fertilization is done by rule-of-thumb, as learned through costly trial and error. Pragmatic observation of nutrient deficiency symptoms of plants, though, is practiced, even by the most experienced.

Some of the more sophisticated techniques now used are: (1) analysis of the whole or specific part (tissue) of the plant, (2) analysis of the soil in which the plant is growing, and (3) biological test of growth response in which plants are grown in pots of the soil in question or on a small area of land.

Since all of these tests have limitations, the experience of the diagnostician is extremely important. The success of the interpretations of the test and recommendations depend upon a knowledge of the area, limitations of the procedure, environmental factors, physical condition of the soils, and disease and insect prevalence. Obtaining a representative sample of the soil to be tested is of prime consideration.

The chemical soil test is one of the most universally used methods, since it permits the detection of deficiencies before the seed is planted and before the plants grow too large for correction.

Phosphorus

Phosphorus is second to nitrogen as a nutrient whose lack most often limits maximum plant growth in arid soils. Phosphorus is applied to soils immediately upon first cultivation from the native condition. Thereafter it is added only about once in two years for most home and urban landscaping.

Leguminous plants such as peas, beans, and alfalfa respond to phosphate fertilizer more frequently than nonlegumes. Under intensive irrigation culture, certain plants regularly demonstrate improved quality due to phosphate fertilizer applications.

Kinds of Phosphate Fertilizers

For convenience, commercial phosphate fertilizers may be listed in five groups, based on characteristics affecting method of general use on alkaline or calcareous soils. The following classification is based primarily on water solubility of the phosphorus compounds (Table 6.2).

Group I: Characterized by high degree of water solubility.

Group II: Partially water soluble, or may undergo a chemical reaction that converts them to a water-soluble form.

Group III: Less water solubility than Group II. Very slowly soluble in soil solution. Solvent action of roots makes these available to plants.

Group IV: Very insoluble in water and soil solution (like rock phosphate).

Group V: Natural organic fertilizers which contain organic phosphorus available to plants indirectly after mineralization as a consequence of decay.

Table 6.2

Examples of Some Common Phosphorus Fertilizer Carriers as Related to the Groupings of Solubility

Name*	Approximate percentages $N-P_2O_5-K_2O$	Relative solubility group
Phosphoric acid	0–54–0	Group I
Ammonium phosphate (several grades)	11–48–0 16–48–0 16–20–0 18–46–0	
Diammonium phosphate	21–53–0	
Superphosphate, triple	0–46–0	
Superphosphate, single	0–20–0	Group II
Calcium metaphosphate	0–61–0	
Ammoniated superphosphate		
Ordinary	3.5–17–0	
Concentrated	5–46–0	
Dicalcium phosphate	0–53–0	Group III
Calcium metaphosphate (special form)	0–61–0	
Basic slag	0–(9–18)–0	Group IV
Rock phosphate	0–(25–37)–0	

*Listed in order of most soluble to least soluble.

All indications are that phosphates in the first three categories are comparable under most conditions if the particles are small and thoroughly mixed in the soil surrounding the roots. Favorable moisture must be available in the area of the fertilizer particles, or roots cannot obtain the phosphorus. Under band placement, however, differences in water solubility may be reflected sharply as differences in availability to plants. The least water soluble phosphates are the least used.

The important difference between water soluble phosphates of Group I and those with limited solubility (Groups II and III) is in the extent of phosphate ions exposed to absorption by roots. The greater the water solubility, the greater are the number of phosphorus ions made available for plant use.

Application

Time. The time of application of phosphate fertilizer is not as critical as with nitrogen fertilizers, because phosphates move very little in soil in the inorganic chemical form. Preplanting applications are most convenient.

Placement. Phosphorus must be in an available chemical form and must be placed so that it is accessible to plant roots, since it moves very little in the soil. It must be present within the root feeding zone of the soil throughout the growth of the plant. This means it must be placed several inches below the surface. Method of application depends on several factors, the most important of which are: (1) type and development (extent) of the plant root system, (2) reaction of fertilizer with the soil, and (3) fertilizer particle size.

Spading under, which will result in thorough incorporation, leaves phosphorus more available to the plant than either applying in bands or broadcasting on the surface. This is especially true when moisture is inadequate near the soil surface, either for the desired chemical reactions or for root contact. Where phosphates are mixed into the root zone, the finer particles are more effective, since they contact more root surfaces than coarser particles.

Broadcast refers to spreading fertilizers evenly on the surface. Adequate moisture for development of feeder roots near the soil surface must be maintained for efficient utilization of broadcast phosphate. This usually cannot be done easily in hot desert climates and is most ineffective with row crops.

Banding usually is confined to row-type planting. It refers to placing a band of fertilizer either below the seed at planting or in a trench several inches deep beside the plant row.

It can be stated categorically that mixing fertilizer in the soil is the best method of application. Relative superiority of a given method, however, depends on factors such as type and age of crop, moisture supply, soil fertility, time of application, phosphorus fixing power of the soil, irrigation pattern, and management.

Selecting a Phosphorus Fertilizer

Phosphate fertilizers may be evaluated according to their *chemical properties, physical properties, effectiveness,* and *cost.*

Chemically, one of the most valued characteristics of phosphate fertilizer is its tendency to dissolve in water and become available to plants by absorption. The physical form of a fertilizer (that is, whether it is finely divided into small particles or granulated for easy flow through placement equipment) is an important consideration. In general, the less soluble the fertilizer, the finer should be its particle size. High phosphorus content is also desirable. Effectiveness of a fertilizer in response to its use in a soil is measured in terms of plant growth and quality.

Cost is most often evaluated on a basis of quantity of available phosphorus per dollar, rather than on how much total bag-weight can be purchased per dollar.

Potassium

As mentioned previously, potassium often is sufficient for plant growth in desert soils. Some growers add small amounts to home gardens and landscaping, just for insurance.

Kinds of Potash Fertilizer

Most commercial potassium fertilizers are water soluble. The more common ones are shown in Table 6.3.

If potash fertilizer is used on arid soils, the least desirable form is potassium chloride. Chlorides can accumulate in desert soils to the extent of causing toxic effects in plants. The nitrate form has an advantage of containing nitrogen, about 13 percent. Generally, potassium finds its way into desert soils as a constituent of mixed or

Table 6.3

Some Potash Fertilizers and Their Content

Fertilizer (chemical name)	N (%)	K_2O (%)	K (%)
Potassium chloride (muriate of potash)	—	60–62	50
Potassium magnesium sulfate	—	22	18
Potassium sulfate	—	50	42
Potassium nitrate	13	45	47

complete fertilizers, for example, 10-20-5 and 10-10-2, containing five and two percent K_2O, respectively.

Natural organic fertilizers also contain potassium in organic combination, which becomes available to plants slowly as the organic matter decomposes, releasing the mineral or inorganic form.

Application

Since potassium moves more readily in the soil than phosphorus, it need not be applied so deeply. Potassium is also held with various degrees of tenacity to the soil particles and is released slowly for plant use.

Selecting a Potassium Fertilizer

Where nitrogen is needed, potassium nitrate may have an advantage because both nitrogen and potassium may be added in one operation. Likewise, where magnesium is required, potassium-magnesium-sulfate provides both potassium and magnesium. As always, cost per unit of elements required can be a factor in selection.

Home gardeners who use an abundance of organic materials such as composts, manure, sludge, grass clippings, and other plant residues will not often need to apply potassium fertilizers to their soils. Occupants of homes built on areas where the topsoil (surface foot, at least) has been removed by excavation should have their soil tested for available potassium before landscaping is undertaken. This is suggested, since there is evidence of a decrease in available potassium with progressive soil depth.

Micronutrients

Except for iron, zinc, copper, and manganese, plant micronutrients are more apt to be present in excesses than deficiencies in soils of the desert Southwest. Moreover, micronutrients such as boron, chlorides, and fluorides may accumulate naturally to a toxic concentration. The chance of deficiencies is lessened by the use of manures and certain mixed fertilizers. Some micronutrients are supplied to soils incidentally with these materials. Manufacturers provide micronutrients in mixed fertilizers for specialty crops, as well as micronutrient fertilizers containing iron, zinc, copper, boron, manganese, and molybdenum. Newer micronutrient fertilizers contain combinations of compatible chelates (chelates are soluble-organic forms of nutrients). The chelates may be added to the soil directly and to irrigation water, sprayed on foliage, or injected into limbs and trunks of trees and shrubs. Directions on containers should be carefully followed, since excesses cause serious damage to plants.

Methods of Application

Soil. Iron and zinc sulfate are commonly used as soil applications to help correct these deficiencies. Plants respond to soil applications more slowly than to foliage applications. The persistence of soluble inorganic iron sulfate in the soil is short-lived in terms of the life of the plant. Inorganic compounds precipitate in calcareous soils as insoluble oxides. However, they may be placed in the bottom of tree holes with compost or manure and an acidifying agent, such as powdered elemental sulfur, to provide some available iron for several seasons. By side-dressing with a sulfur + manure + iron sulfate mixture, lime-induced chlorosis (yellowing) in some plants has been corrected. The side-dressed mixture must be worked into the soil at the side of the plant or row but not so that it will contact the roots excessively.

Correction of chlorosis (yellow-colored foliage) in trees and shrubs is much more difficult than it is in garden plants and annuals. The sulfur-organic-iron sulfate mixture should be worked one to two feet deep into the soil. This is difficult to do without injuring roots. This danger may be overcome in part by drilling holes and placing the material where the roots are least dense.

The sulfates of zinc, copper, and manganese remain available for plant uptake over a much longer period of time than does iron sulfate.

Chelated micronutrients do not become insoluble and precipitate in calcareous soils as readily as the highly ionized inorganic salts ("chelate" means to hold, as in a bird's claw). Some may remain available to plants as long as a year after application. Moreover, acid-loving broadleaf evergreens which become chlorotic readily in desert soils and do not respond to inorganic chemicals can be restored by use of chelates suitable for alkaline-calcareous soils. Because of the greater availability of chelates to plants, much less chelate than inorganic sulfate salts is required, although correction of chlorosis due to iron deficiency in calcareous soils is more difficult than in acid soils.

Injections into limbs and trunks. Many trees respond to micronutrient applications placed directly in holes drilled into the limbs and trunks. Injections of iron sulfate into the trunks of some trees will last for several years, and iron chlorosis can be kept under control by repeating the treatment every few years. Drilling holes (¼ inch) in tree trunks less than four to five inches in diameter may weaken the trees too much for this type of application to be practicable. Not all trees respond to this treatment. One is therefore obliged to experiment carefully to determine the best method of injection.

Iron citrate is claimed to be a better source of iron than iron sulfate, and chelates appear to be the most effective of all. No general recommendation seems to suit all cases. Driving iron nails and zinc-coated nails into trees is an old remedy sometimes still practiced, although unreliable and erratic in effectiveness.

Foliage applications of micronutrients are very popular, even though success in this practice is not always possible, since the leaves of certain plants resist wetting. Where they do not, results appear quickly and correction can be as complete as the extent, timing, and frequency of the spraying. Chelated micronutrients are common constituents of foliage sprays. Where large areas of plants are involved and costs must be kept to a minimum, the iron-sulfate foliage spray shown in Table 6.4 may be used with good results on many plants. Foliage sprays have the disadvantage of correcting only the plant part they touch. New growth requires new spraying if factors causing the deficiency persist.

Table 6.4

Iron Sulfate Foliage Spray*

Materials	Amcunt
1. Iron sulfate in distilled water (if manganese or zinc are required, add 0.5% manganese sulfate or 0.25% zinc sulfate)	1%
2. Detergent (non-phosphate)	Trace
3. Citric acid (use with caution and only when spreading on foliage is seriously inhibited)	1%

*After W. T. McGeorge (1954). A new spray for the cure of lime-induced chlorosis. *Better crops with plant food.* Am. Inst. Potash, Wash., D.C.

When the micronutrient solution is used for large acreages, citric acid should be left out because it readily corrodes the metal parts of the spray equipment. Citric acid may be used for the home garden where employment of spray equipment is not prolonged. Citric acid helps to spread the iron sulfate and prevents it from precipitating on the foliage.

For spray application, rates of one, two, three and four pounds of iron chelate per 100 gallons of water (or proportionately less on a gallon basis) have been used with success. The foliage must be soaked. *The directions on the container should be followed.* Using more than is recommended may result in toxicity problems and the shedding of leaves and blooms.

In addition to chelates of iron, chelates of zinc, manganese, copper, and magnesium are available. Seldom do ornamentals require the latter micronutrients in desert soils of the Southwest.

The use of chelates, or other micronutrient treatment, should not be substituted for proper soil and water management practices to permanently correct micronutrient deficiencies.

Conversion Factors

To convert the weight of different fertilizers into commonly used volumes, the factors in Table 6.5 will be helpful in relating ounces to teaspoons, cups, or pints. The amount of inorganic and organic fertilizers to add to a given area of soil in these measurements is shown in Table 6.6.

Table 6.5

Approximate Volume-Weights of Different Fertilizers

Fertilizer	Teaspoonful (heaping)	Standard cup (level)	Pint
	Ounces	Ounces	Ounces
Ammophos 16–20; Ammophos 11–48	⅔	7.0	14
Ammonium sulfate; sodium nitrate; ammonium nitrate	¾	7.5	15
Superphosphates, treble-superphosphates	¾	8.0	16
Mixed fertilizers	1	8.5	17

Table 6.6

Amount of Fertilizer to Add to a Given Area of Soil

Fertilizer material	Incorporated during seedbed preparation*			Maintenance application as needed*		
Nitrogen	**1 sq. ft. teaspoons**	**9 sq. ft. tablespoons**	**100 sq. ft. cups**	**1 sq. ft. teaspoons**	**9 sq. ft. tablespoons**	**100 sq. ft. cups**
Calcium nitrate	¾–1½	2½–4	1¾–2¾	½	1¾	1¼
Ammonium sulfate	¾–1¼	2¼–3¾	1½–2½	½	1½	1
Ammonium nitrate	½–1	1½–2¾	1¼–2	¼	1	¾
Urea	½–¾	1¼–2	¾–1½	¼	¾	½
Phosphates						
Superphosphate	1–2	2¾–5½	2–4	Phosphates to be applied as maximum every other year		
Concentrated superphosphate	½–¾	1¼–2¼	¾–1½			
Ammonium phosphate	½–¾	1¼–2	¾–1½			
Organics						
Manure or compost	¼–½ lb	2–4 lb	25–50 lb	Add as needed		
Straw, leaves, or peat moss	¼ lb + ½ t N fertilizer	2 lb + 1 T N fertilizer	25 lb + 1 C N fertilizer	Add as needed		

*Measuring spoons and cup should be no more than level full.

Fertilizer recommendations more often are made on a large scale basis, such as pounds per acre. Therefore, to convert this information into values useful to the home gardener, who plants on a smaller area, Table 6.7 was prepared. Rates of applications can be made on a 50-foot row basis, or a broadcast square-foot area basis. Rates of application are shown in terms of ounces.

Data in Table 6.8 are presented for converting the usual N - P_2O_5 - K_2O basis to a single elemental basis of N-P-K (for example, phosphorus or potassium instead of the oxide forms).

Table 6.7

Weights of Fertilizer Calculated for Small Areas

Rate per Acre lb*	Corresponding amounts of fertilizer for row plantings: Approximate amounts of fertilizer per 50 ft of row in widths of:						Equivalent weights per 100 sq ft
	12 in (oz)	18 in (oz)	24 in (oz)	30 in (oz)	36 in (oz)	42 in (oz)	(oz)
100	2	3	4	5	6	7	4
150	3	4.5	6	7.5	9	10.5	6
200	4	6	8	10	12	14	8
300	6	9	12	15	18	21	12
400	8	12	16	20	24	28	16
500	10	15	20	25	30	35	20
600	12	18	24	30	36	42	24

*1 pound = 16 ounces.

Table 6.8

Fertilizer Elemental Conversion Factors

Percent or pounds		Multiply by factor		Converted to percent or pounds
Conversion to Element				
P_2O_5	×	0.44	=	P
K_2O	×	0.83	=	K
Conversion to Oxide				
P	×	2.29	=	P_2O_5
K	×	1.20	=	K_2O

7. Some Soils Need Conditioning

Soil-conditioning is more common to the arid and semiarid areas of the West than to humid climates. Some conditioning methods originated as an aid in soil reclamation, whereas others were developed to eliminate or counteract poor water movement in soils.

Conditioning is required of soils which have excesses of certain harmful salts, poor internal drainage, heavy clays, compacted layers, hardpans, crusting problems, or some obstruction. Almost always it involves improvement of some physical property of the soil, such as structure, for more favorable plant growth. Soil-conditioning may promote chemical replacement of sodium with calcium, which process flocculates the soil particles and allows freer water movement and aeration. Soil-conditioning may involve improvement surrounding the seed such that it germinates more readily and the seedling is better established. Sand, vermiculite, perlite (exploded glass), sawdust, or well-rotted manure exemplify materials which may be added with the seed or to the seed area of soils which contain lime and tend to crust and bake in the desert heat. Conditioning of the soil may involve mechanical manipulation such as deep tilling to break up texture-stratified soil, hardpans, and compacted layers. In any event, the purpose is to make the soil more favorable for water movement and aeration, with consequent improvement in root growth and penetration. The purpose of conditioning is not to add nutrients, though it may affect the nutrient status of the soil incidentally or indirectly.

Soil Amendments

The term *soil amendment* is defined in the Arizona State Fertilizer Materials Act as follows:

Soil amendments, for example, shall include: Wood charcoal, pumice, perlite, expanded vermiculite, sintered shale, diatomite and clay; if sold as such with no claim for chemical constituents and intended for use solely because of their physical nature. Organic materials such as organic polyelectrolytes shall be considered as soil amendments if they affect the physical properties of soils. Mixtures sold as potting media without further beneficial claims are soil amendments.

The term *soil conditioner* was avoided in the definition above because it has been so badly misused as to have no clear meaning to many people. It commonly refers to a heterogeneous host of materials, organic as well as inorganic, which may influence the biological, chemical, and physical condition of the soil, though usually it refers to materials which primarily alter or change the soil's physical properties. Many commercial materials of doubtful value have been claimed as soil conditioners. They appear on the market through vigorous sales promotion with little or no suitable evaluation having been made to substantiate promoters' claims. These questionable materials do not survive but in their transient appearance they confuse the understanding of and damage the general concept of soil conditioners.

Soil amendments are academic orphans in the sense that they have received little scholarly attention except in connection with reclamation and seldom appear as a topic in texts on soils and fertilizers. This is probably due, in part at least, to the fact that many of them escape precise definition and evaluation. Reclamation, a field which is highly technical and associated with arid lands, receives slight attention in textbooks. The few books which contain information about reclamation are aimed at the technical reader.

The term *soil-conditioning* is used in this text to mean the addition of any substance or combination of substances put on or into the soil, as well as mechanical manipulations of the soil, which influence its physical properties in such a way as to benefit plant growth. This concept does not rule out materials or tilling that may either directly or indirectly influence the chemical properties of the soil or the nutritional status of plants growing in the soil. Even this definition is not wholly satisfactory since (a) certain materials may, under some peculiar circumstance, but not under others, favorably influence soil properties, and (b) experimentation to fully evaluate an alleged soil-conditioning effect is very expensive and time-consuming and, therefore, often has not been systematically undertaken.

Functions of Soil Amendments

The productivity of the soil is critically influenced by its *structure*, or arrangement of the soil particles in a profile. The air, moisture, and temperature relationships in a soil are controlled largely by structure. Structure also affects the uptake of essential nutrients and absorption of salts by roots.

The function of a soil-conditioning material is evaluated in terms of: (1) cementing capacity, which is manifested in the *number* and *sizes* of soil particles it is capable of holding together in desirable water-stable aggregates, (2) influence on soil moisture movement, retention, and availability, (3) influence on soil air space and temperature, (4) influence on nutrient availability, (5) influence on the control of undesirable salt accumulation, (6) influence on the microbial activity, and (7) long persistence in the soil. No single soil amendment will fulfill all these objectives. Most of the soil amendments listed below may be found at a nursery or fertilizer store.

Classes of Soil Amendments

As a matter of convenience, soil amendments and conditioners are grouped into general categories of like materials, as follows:

Group I includes *inorganic chemicals* such as gypsum, sulfur, sulfuric acid, polysulfides, and iron sulfates. These affect chemical reactions and provide soluble calcium for replacing sodium and/or coagulating the soil clay particles into blocks, as in water clarification or purification using salt.

Group II includes *inert materials* such as sand, gravel, ground rock, exploded silicates (perlite), coal ashes, vermiculite, and glass wool. These materials, if added in sufficient quantity in band or hill applications, can change the physical condition of the soil by altering the texture and structure around the seed. The extent of change depends, of course, on the quantity added, and depth and distribution of application, i.e., how much material is added in how much soil. Thus, the effectiveness of these materials is dependent upon an understanding of the concept of *dilution* and the amount of material necessary to permit measurable effects from dilution. For example, an acre six-inches of air-dry soil weighs about 2,000,000 pounds. This is a large amount of soil to be altered significantly in texture for the practical purpose of growing plants. Therefore, if these Group II materials are banded or placed in hills, i.e., placed with seeds, making a high concentration per unit area or weight of soil, there is more opportunity for them to be of practical benefit to the plants at a minimum cost than if mixed with all the soil to a depth of six inches and thus be highly diluted.

Group III includes *all materials of plant and animal origin.* This group primarily affects the soil structure. The effect depends largely on the rate of decomposition of organic materials present. Peat, sawdust, bark, straw, woody stems, vines, branches, and roots decompose relatively slowly and therefore alter soil structure directly by holding soil particles apart, thereby enlarging the pore space and making the soil loose. Such materials as manures, composts, sewage sludge, dried blood, fish residues, guano, grass clippings, and plant tops decompose more rapidly. They affect the soil more indirectly by providing energy for the creation (by microorganisms) of bonding substances that aid in the development of a desirable crumblike structure. The more food supplied to soil microorganisms, the more biological gums are produced. The soil stabilizing action of gums produced by microorganisms is well-known. Addition of microorganisms to the soil is *not* necessary to improve this production; only an available food source is needed.

Group IV includes *synthetic organic compounds*, such as polyelectrolytes or hydrolyzed polyacrylonitrile (HPAN), vinyl-acetate-maleic acid compound (VAMA), and wetting agents. HPAN and VAMA cement soil aggregates together, preventing their ready dissolution and dispersion into a structureless mud puddle. They possess no property for *making* soil structure. They *stabilize structure* already formed.

Because some soils in arid regions have been found to repel water, *wetting agents* (surfactants) have been tested as possible aids in increased infiltration of water. Some success under certain conditions has been achieved, and several "brand name" wetting agents are available.

Group V includes *soil inoculants, microbial cultures, enzymes, vitamins, and trace-growth substances.* The only soil inoculants or cultures of microorganisms that have proven economic value to the gardener are cultures added to specific legume species (sweet peas, lupines, beans, etc.) to produce nitrogen-fixing root nodules. Legume bacteria (*Rhizobia*), if inoculated to the proper host seed, aid the legume plant by forming nodules on roots in which atmospheric nitrogen is made available to the plant.

Enzyme preparations and vitamins are of no proven value when applied to soil to enhance plant growth.

Group VI consists of *mechanical soil treatment*. Soil-conditioning involves more than adding soil amendments to the soil. Certain mechanical treatments can improve the physical properties of the soil, particularly structure and pore space.

1. *Tillage practices* condition the soil by providing a more favorable environment for plant roots. Compact layers and surface crusts are broken up. Often the soil becomes so compacted around perennials that the plants must be removed and the soil again deep-tilled.

2. *Seedbed preparation*. One of the most important single operations affecting the favorable growth of plants around the home is the construction of a good seedbed.

3. *Dense cover crops* such as ryegrass, Papago peas, clover, or other green manure crop may be grown effectively to aid in making a mellow soil. Mass plantings of annuals often directly improve the physical condition of the soil by their penetrating root action. A good plan for annual flower beds is to provide a rotation system of dense growth at least once every other year. Bulbs may effectively be mixed with flowering annuals, particularly when the flowering periods complement each other, that is, do not take place exactly at the same time. Addition of organic residues is made when the seedbed is prepared.

Use of Specific Soil Amendments

Group I

Gypsum. Calcium sulfate (gypsum) is applied to desert soil to improve the physical condition and not the nutritional content (Fig. 7.1). Desert soils are already abundantly supplied with available calcium and sulfur as plant nutrients. Moreover, irrigation waters contain appreciable amounts of calcium and sulfur (as sulfate) available for plant use. Gypsum functions specifically in the mass *replacement* of undesirable concentrations of sodium in soils, and not as a fertilizer. This is accomplished by adding it as a powder

Fig. 7.1. Gypsum spreading prior to reclamation leaching of new lands.

to the soil and flooding with water two or three times to soak the soil to a depth of about five feet. In this situation, a large mass of soluble calcium replaces sodium, the undesirable salt. Gypsum is so slowly soluble that large quantities must be supplied in addition to that in the soil to accomplish replacement and achieve reclamation.

Gypsum also counteracts sluggish water movement in soil dispersed as a result of excess removal of salts during irrigation or after a heavy rain as sometimes occurs in summer. The calcium ions favor flocculation of the soil clay, which provides more pore space for water and air movement. Not all soils benefit physically by gypsum use. If water penetrates soil readily, gypsum or any other amendment probably will be of little value physically. Usually, home gardeners and landscapers do not use gypsum in appreciable quantities as does commercial agriculture, although there are times when the soil would benefit by its addition.

When required, only high quality gypsum should be used, and it should be very finely ground (300 mesh). The Arizona Fertilizer Materials Act stipulates that, "the percentage of calcium sulfate contained therein be indicated clearly on each source of gypsum or agricultural mineral, the principal constituent of which is calcium sulfate."

Gypsum is best added to the soil directly and mixed in before irrigating. The more intimate the association between soil particles and gypsum, the greater is its effectiveness. Gypsum does not burn plants.

Sulfur. Certain chemicals used agriculturally react with calcareous soils to produce gypsum. Some of the most widely used chemicals are sulfur, sulfuric acid, sulfur dioxide gas, calcium polysulfide, and iron sulfate. These substances produce about the same soil-conditioning effects as gypsum if applied in quantities which would create an equivalent amount of gypsum (see Table 7.1).

Table 7.1

Amounts of Various Amendments Required to Supply 1000 Pounds of Soluble Calcium when Applied Under the Proper Soil Conditions

Amendment	Purity	Pounds required to supply 1000 pounds of soluble calcium
Gypsum ($CaSO_4 \times 2H_2O$)	100	4300
Sulfur (S)	100	800
Sulfuric acid (H_2SO_4)	95	2600
Iron sulfate ($FeSO_4 \times 7H_2O$)	100	6950

Sulfur is oxidized to sulfuric acid by the action of soil microorganisms. The acid then reacts with lime, yielding gypsum and carbon dioxide gas. Sulfur is used to reclaim sodium soils, prevent crusting, and enhance the availability of certain plant nutrients in soil. With respect to the latter point, sulfur has been shown to assist in the correction of lime-induced chlorosis by making soil iron more soluble.

Sulfur is added to tree and shrub holes before planting and worked into the soil of seed beds. The action of microorganisms on sulfur is slow. Finely powdered agricultural sulfur may take three to four weeks to oxidize completely, depending on how well it is mixed with the soil. Oxidation to sulfuric acid will not take place in dry soil and during cold winter months. Conditions that favor high microbial activity also favor a more rapid rate of sulfur oxidation.

Large pieces of flake-sulfur oxidize very slowly and may remain in the soil for years. This is due to the formation of sulfuric acid on the surface of the flakes in such high concentration that the buildup of acidity inhibits further oxidation by microorganisms.

Sulfuric acid. Sulfuric acid (H_2SO_4) reacts with lime in the soil to produce gypsum ($CaSO_4 \cdot 2H_2O$). In desert soils the reaction is immediate, since it does not require the action of microorganisms before it can react with the soil lime to make gypsum.

Battery acid is sulfuric acid. To understand its corrosiveness, one need only visualize its effect on battery cables, holders, and metal parts of the automobile. Because of danger in handling, it is seldom used around the home. In fact, only experienced handlers are permitted to apply sulfuric acid to soils.

Sulfur dioxide. Sulfur dioxide (SO_2) is a colorless, noninflammable, corrosive gas. It is not used around the home, but is commercially applied to irrigation water used in agriculture. In water it forms sulfurous acid which oxidizes to sulfuric acid upon contact with the soil.

Polysulfides. These are complex mixtures of sulfides, sulfates, thiosulfates, and molecular sulfur. The polysulfide most commonly used on desert soils is the calcium form. It is a highly alkaline, brown liquid. The most convenient way to apply is to put it in the irrigation water. When added to water, very fine, colloidal elemental sulfur precipitates, giving a white, cloudy appearance. Commercial products contain about 23 percent sulfur and six percent calcium. Benefits to the soil are about the same as with powdered sulfur. It is not corrosive to concrete or steel, as is sulfuric acid. Polysulfides often are used to help correct lime-induced iron chlorosis in home soils and public landscapings. Polysulfides condition the soil also through the indirect formation of gypsum.

Mine residues. Mine tailings containing *sulfides* of iron, manganese, copper, and zinc accumulate in large piles from the Southwest's mining enterprises. *Ferrous sulfide* (pyrite) is the most prominent sulfide present. Other iron-sulfur compounds are double sulfides (pyrrhotite) or multi-sulfides (polysulfide). Some of these wastes may be used directly as soil conditioners; others require strong treatment such as with sulfuric acid for them to be effective.

Pyrite "soil conditioners" have appeared on the agricultural market for many years, and even in the early 1970s a few could be found for sale, despite their very insoluble nature and tendency to be slowly oxidized. The rate of oxidation by microorganisms is so sluggish that pyrites are ineffective for soil conditioning or for supplying plants with significant amounts of available iron.

Pyrrhotite, which represents another group of iron-sulfur compounds, is oxidized quite rapidly by native soil organisms, giving it a soil-conditioning character. Copiapite, a naturally occurring oxidation product of iron pyrite, also appears to have soil-conditioning possibilities. In fact, the effects of pyrrhotite and copiapite on the soil are similar to those of iron sulfate. The concentration of these compounds in raw mine tailings is small.

Iron sulfates, ferric or ferrous sulfate, have been used to aid in preventing the crusting of soils. Ferrous sulfate is a byproduct of mining. Iron sulfates rapidly react with moist soil to form gypsum and insoluble hydroxides and oxides of iron. Ferr*ous* sulfate is a clear gray-green salt that turns to ferr*ic* sulfate, a gray powder, on exposure to air. This happens fairly rapidly in containers around the home. Often calls are received asking if it is suitable to use after it changes form. It is. Applications of iron sulfate may be made directly to the soil and worked into the top few inches. Experimental results, using iron sulfate as a soil conditioner, have not been encouraging. Iron sulfate reacts with so many plant nutrients in the soil to form insoluble compounds that its frequent use in large quantities is not recommended.

The choice of a soil amendment in Group I will depend on: (1) nature of the soil type with respect to the total salt and sodium content, (2) availability of material and ease of handling, (3) time required for the amendment to react in the soil, (4) cost of amendment per unit of calcium it can supply or which can be released from lime in the soil, and (5) effectiveness in improving the physical properties of the soil as related to favorable influence on plant growth.

Gypsum is the only material which supplies calcium directly. All the other materials require that indigenous lime be present in the soil. Gypsum, sulfuric acid, and iron sulfate react relatively rapidly. Sulfur needs to be oxidized by microorganisms in the soil before it can provide soluble calcium; therefore, it acts more slowly. According to calculations in Table 6.1, more than one and a half times as much iron sulfate as gypsum is required to provide an equal quantity of calcium. However, if gypsum or calcium is needed by a soil, and it must be supplied through the reaction of iron sulfate, serious problems can develop as a result of large applications of iron salts.

Group II

Sand, gravel, stone, and ground rock. These materials may influence plant growth by changing the soil texture and pore space. Unless the area is extremely small, any hope of changing the texture sufficiently to affect plant growth is out of the question unless many tons of materials are added. For example, to increase the sand content of an acre-foot of soil as much as 10 percent by weight, which is far short of being enough to change a soil's physical makeup appreciably, 400,000 pounds or 200 tons of sand would be required.

On the other hand, rock mulches are being used effectively as "surface armor" in preventing soil and water erosion on slopes (Fig. 7.2). Gravel and crushed rock often replace turf for home yards and patios. Planting strips and small patios in shopping and apartment areas are well-adapted to the use of gravel and rock mulches. They are unusually practical because of their ease of maintenance. The use of large stones and natural rocks for mulching residential

Fig. 7.2. Rock mulch on sloping land prevents soil and water erosion and can enhance the beauty of the patio (photo by Clair Cameron).

Fig. 7.3. Applying asphalt mulch over the seed row in desert soil.

areas is very popular. Large rocks placed in highway dividing areas and medians keep cars from crossing at hazardous places. The maintenance of the medians is minimal, views are not obstructed, and the rock mulch effect can be aesthetically pleasing.

Exploded silicates. Exploded silicates (perlite) may be applied to soil surfaces to *reduce* the temperature by means of the white color. Other than this use, or to mulch seedbeds, these materials are very costly in amounts sufficient to change significantly the physical properties of a soil. However, perlite may be used effectively and economically in greenhouse and potting soils. Generally, the white particles in floristic pots are perlite or a similar exploded silicate.

Coal ash, coke, carbon black. Black granules, dusts, and powders of wood ash, coal ash, coke, and carbon black have been applied to soil surfaces to *increase* their temperature during cold months. The increased temperature of the soil beneath the banded materials of row crops and vegetables gives the young seedlings a head start in winter or early spring in the mild climates of the desert Southwest. These materials have not been put to widespread use as yet because of handling problems. Erosion of these mulches by rain handicaps their use in some instances. Black plastic film mulch has been used successfully as a conditioner for the purpose of warming up the soil early in the season, preventing crusting, keeping weeds under control when plants are small, and conserving soil moisture.

Asphalt or petroleum mulch. Petroleum mulches have been shown to be effective in moisture retention and increasing the soil temperature, resulting in earlier seedling emergence, better stands, and earlier yields (see Fig. 7.3). The energy crisis in 1973 brought an end to the use of asphalt for such limited uses as mulch.

Vermiculite. Vermiculite (exploded mica) is effective in improving the stand and seedling-survival of plants from small seeds. Placed along with lettuce seed, for example, it combats soil crusting and improves germination and emergence of young plants. Vermiculite is successful when placed on the soil surface in pots in the greenhouse as an aid for the control of damp-off disease and crusting.

Group III

Peat, sawdust, and bark. Woody materials such as peat moss, sawdust, and bark residues can be effective amendments for desert soils. All aid considerably in preserving the moisture of the soil and assist in the control of alkalinity. Because of their low nitrogen and high carbon content, i.e., wide C/N ratio, nitrogen should be mixed with them or added to the soil. These woody materials are best used around the home in planter boxes and in mulches at the base of ornamentals and fruit trees, since their cost practically eliminates their application to large acreage of agricultural crops. However, sawdust compost has been used successfully for commercially grown roses where fairly large acreages were involved. Sawdust also makes an excellent plant container mix in a ratio of three parts sawdust to one part soil by volume. Sawdust from pine, fir, and alder has a favorable effect on the physical properties of soil when mixed to a depth of six to eight inches. Its value is improved by fortifying with phosphorus as well as with nitrogen fertilizers.

Bark and sawdust both excel as mulches placed on the surface of the soil, whereas peat moss has proved valuable when mixed into the soil. Peat moss is especially useful in soil planted to "acid-loving" ornamentals to help control excess alkalinity and provide a more favorable rooting medium. Bark is very slow to decompose. It is primarily a physical dilutent, since it contributes little to microbial activity or the nutritional needs of plants until it has been in the soil for a long time.

Straw, leaves, grass clippings, and woody plant residues. For centuries, plant residues have been the backbone of man's food production economy. They maintain the physical, biological, and nutritional status of the soil. As with peat, sawdust, and bark, and with all carbonaceous materials generally, it is imperative to add extra commercial nitrogen such as ammonium sulfate to the soil when plant residues are added.

Plant residues benefit growth in many ways, and condition the soil as well as favorably affecting plant nutrition. Some of the soil-conditioning effects that may be attributed to plant residues are: (1) promotion of a favorable soil structure, (2) promotion of favorable moisture relation, water movement, and water conservation, (3) provision of a good tilth, and thereby ease of working and good root penetration, (4) contribution to ion exchange colloids, thus aiding in plant nutrition by holding nutritional elements against loss through leaching, and (5) addition of humus which acts as a slowly available food source for soil organisms to produce structure-forming substances.

Further discussion on straw and plant residues for soil improvement appears in Chapter 4.

Manures are well-known for their favorable effects in the soil. They are particularly valuable to home and landscape gardeners. There is some question regarding their role in commercial agriculture. The cost of transporting animal residues from feed pens and spreading them on the land has been questioned as a matter of economic return for dollars spent. Commercial fertilizers provide the necessary plant nutrients at a lower cost. Therefore, to be an effective competitor, manures must supply additional benefits. This they do; but are they enough? When applied to soils that need conditioning, high quality manures benefit plants beyond what they can supply nutritionally by improving the physical condition of the soil.

Manures differ in original nutrient composition (Table 7.2). The content of N, P, and K reported in this table is somewhat higher than is found in manures from stockpiles and the old barnyard com-

Table 7.2

Total Nitrogen, Phosphorus and Potassium in Manure from Different Animals

Whole manure* (no straw or bedding)	N (%)	P (%)	P_2O_5 (%)	K (%)	K_2O (%)
		moisture-free basis			
Horse	3.2	0.48	1.1	2.1	2.5
Cow	4.3	0.48	1.1	2.7	3.2
Sheep	3.1	0.48	1.1	2.9	3.5
Swine	3.8	1.14	2.6	2.6	3.1
Hens	2.2	0.80	1.8	5.6	6.7

*Whole manure includes both feces and urine.

post heap. Soil, sand, straw, and bedding residues become a part of stockpile manures. Leaching away of valuable nutrients occurs where stockpiles are exposed to the weather. Some chemical characteristics of manures from five feedlots and a dairy in Arizona are given in Table 7.3, in which it will be noticed that the sodium content is quite variable and averages 0.85 percent (or 8,500 ppm), although, as in most western states, feedlot manure potassium salts exceed those of sodium.

The high content of soluble salts in feedlot and dairy manures is a serious problem in manures marketed for garden use. In eight samples of steer manure from four feedlots in the Phoenix area, soluble salts varied from 4.42 to 8.73 percent. Manure of this composition will add 88 to 175 pounds of soluble salt per ton applied. These manures do not always represent a potential benefit to plants and may even be a detriment to growth, depending on the specific salt tolerance of the plant.

Favorable responses in plants to animal manure additions depend on a number of factors, which may be summarized as follows: (1) *Quality*. Insist on high quality manure, free of weed seeds, sand, and dirt, and low in total salts and sodium. (2) *Quantity*. Apply no less than one-half pound per square foot every year for best results. (3) *Age of manure*. Aging a certain amount in the stockpile before spreading on the land is preferred. Uniformity of aging is also important. (4) *Time of application*. Manures should be applied well in advance of planting. A month in advance may be minimal. (5) *Method of application*. Apply as a broadcast over the land and spade into the soil for most effectiveness. (6) *Place of application*. Manures are not effective when used on high salt spots and on soils where downward penetration of water cannot take place throughout the root zone because of the sodium content.

Bat guano deposits also vary widely in nutrient content. This well-known and respected fertilizer of long history ranges from nine to 14 percent nitrogen in the richest deposits and from 1.5 to 4.5 in the poorest (Table 7.4). Flooding and silting of guano deposits during great climatic changes in history are responsible for their characteristic stratification in quality. Those deposits richest in plant nutrients, as reported in 1915 (Table 6.4), long since have been mined. The remaining guano contains low values of nitrogen and phosphorus. Its commercial importance, at present market value,

Table 7.3

Composition of One Dairy Manure and Five Feedlot Manures*

Sample	Dry weight (%)	Organic[1] matter (%)	Sand[2] (%)	Nitrogen[3] (%)	Phosphorus[4] (P_2O_5) (%)	Phosphorus[4] (P_2O_5) (lb)	Soluble[5] salts (%)	Calcium & magnesium (%)	Sodium (%)	Potassium (%)
Dairy	80	49	26	1.6	1.2	18	8.9	0.15	1.1	2.3
Feedlot A	48	62	22	1.8	1.3	13	10.6	.08	0.6	3.8
Feedlot B	69	53	29	1.9	0.8	12	8.0	.27	0.8	2.2
Feedlot C	75	40	38	1.6	1.4	20	9.5	.13	1.2	2.4
Feedlot D	80	44	41	1.6	1.2	19	4.2	.06	0.5	1.2
Feedlot E	76	38	47	1.4	1.0	15	7.2	.10	0.8	2.0
Average	71	48	34	1.65	1.14	16.3	8.1	0.13	0.85	2.3

[1]Oven-dry weight less weight of ash.
[2]Ash not soluble in hydrochloric acid.
[3]Total, not including ammonia, etc. lost upon drying at 75C.
[4]Total, by ashing 16 hours at 45C.
[5]Determined in 1:20 manure:water extract.
*From J. L. Abbott. 1968. *Use animal manure effectively.* Univ. Ariz. Agr. Expt. Sta. and Coop. Ext. Serv. Bul. A-55.

Table 7.4

Nitrogen and Phosphorus Content of Some Commercial Bat Guanos from Historic Arizona Caves Sampled in (about) 1915*

Source	Nitrogen N (%)	Ammonia NH_3 (%)	Phosphorus P_2O_5 (%)
Pima County	12.7	15.4	4.1
	14.7	17.9	—
	9.2	11.2	5.9
Maricopa County	8.7	10.5	—
	14.6	17.7	—
Gila County	10.3	12.6	—
Cochise County	9.7	11.8	2.6
	3.6	4.4	2.5
	2.2	2.7	2.2
	1.9	2.3	1.0

*(A. E. Vinson, 1915. *Timely Hints for Farmers.* Univ. Ariz. Agr. Expt. Sta. Bul. 109.)

does not warrant mining and transport. When high-analyses commercial fertilizers were not available and plant production went by rule-of-thumb, growers took advantage of the fertilizing value of bat guano. Even at the high cost per nutrient content, some gardeners still use it as a soil conditioner.

Guano has a very characteristic earthy odor. It contains droppings (some still visible in form), insects, and bat skeletons; even bat fur can be identified at times. Unfortunately, the great stores of rich deposits have been nearly exhausted in the desert Southwest, including the guano accumulated from droppings of fish-eating birds. History, however, has not closed the book on the bat. Secluded caves in the desert Southwest still house bats and their deposits are accumulating in the arid environment, albeit ever so slowly.

Compost. Homemade composts provide an important source of material for the replenishment of soil humus in gardens and residential planting. Compared with commercial chemical fertilizers, they are low in plant nutrients and therefore are improved considerably when fortified with chemical fertilizers, particularly nitrogen (see Chapter 4). The danger of perpetuating plant diseases through home composts must always be evaluated when considering their use.

Compost, as a byproduct of municipal waste disposal, is one of the valuable resources available to agriculture. Although this waste has been ignored in the United States, the advantages of composting municipal refuse (mostly home garbage) are recognized in many foreign countries. The Netherlands, for example, composts over 35 percent of its city refuse. Municipal compost programs are well known in Europe and other countries such as Israel, Japan, and China.

Composting of municipal refuse (city and residential solid waste) is generally considered by the researcher in the field of waste disposal not only to be technically feasible, but also essential to the health and welfare of society. Some of the advantages of composting municipal refuse are: (1) It supplies a source of organic matter for maintaining and building seriously depleted supplies of humus in soils. (2) It improves the growth and vigor of crops and home garden plants. (3) It provides a means for reclaiming certain valuable materials from refuse. (4) It also provides an effective and sanitary method of refuse disposal.

Composts prepared in a biodigester at elevated temperatures by the aerobic microbial decomposition process have been shown to be good sources of organic matter for soils. Pathogenic bacteria and fungi are killed by the heat of digestion. Weed seeds, insects, and insect eggs are destroyed. Most municipal compost is windrowed and allowed to continue to age in a pile, just as in the home compost pile. Some composts may be screened and made into pellets, or left "as is" for bulk uses.

Compost from municipal refuse has certain advantages over other organic materials for maintaining soil humus and improving plant growth. It is low in harmful salts, low in total salts, free from plant diseases and pathogenic organisms harmful to man and animal, free of weed seeds, free of excessive dust, higher in certain plant nutrients than most organic soil conditioners, does not give off objectionable odors, and decays fairly rapidly in the soil.

Pelleted compost has additional advantages over and above those just listed. Pellets are convenient and easy to handle, are cleaner to handle and cleaner around the private home after applying, are practically odorless, and will not blow away as readily by wind action. A choice of different-sized pellets may be selected for specific purposes to suit the fancy of the gardener.

The fine, granular-sized pellets, because of their relatively heavy density and small size, move down through turf grass until they contact the soil, while the larger pellets are ideal for flower beds and mulching.

Both compost and manures may make excellent banded mulches by placing the material loose in a band with the seed as is suggested for the more inert materials of vermiculite, sand, perlite, or asphalt.

Sludge. There are two quite different kinds of sewage-disposal sludges, depending on the system from which they come. *Digested sludge* is produced by anaerobic treatment and has low fertilizing quality compared with *activated sludge*, which is produced by aerobic treatment. Activated sludges are used as fertilizers for lawns and for turfs of golf courses. Digested sludges are used as mulches and in organic incorporation around shrubs, trees, and perennials in home gardens. Digested sludges should be heat treated to assure destruction of disease organisms before use in gardens where vegetables are grown for eating raw. Sludges have fair water-holding capacity and improve water infiltration, and are slow to decompose. They promote vigorous microbial activity and heavy root growth. Sludges are used much like manures. Some have the disadvantage of producing objectionable odors. Sludges are mixed with other soil conditioners with good results. Municipal waste compost is improved by mixing with sludges.

Fish wastes. Waste effluents from the fish industry, containing 90 to 99 percent water, have been on and off the market for many years. Their conditioning value is based on their organic matter content, which stimulates soil microbial activity. The effect of fish waste effluents is attributable to the amount of organic matter they supply to the soil microflora, and not to strange and new kinds of microorganisms they add. These wastes have a limited value for plant production, as do any other effluents of low organic content. Obviously, the cost involved in freighting material containing such a high proportion of water makes the cost extremely high for the amount of organic material provided.

Organic meals and solution concentrates. Fish and blood meal are another story. Both meals have soil-conditioning properties. Their value, however, is greatest with respect to their plant nutrient content — namely, nitrogen and phosphorus. As organic fertilizers they are excellent, but the cost per unit of nutrient element is high compared with inorganic chemical fertilizers.

Group IV

Synthetic organic polyelectrolytes. The organic chemicals, hydrolyzed polyacrylonitrile (HPAN or Krillium) and vinyl acetate maleic acid compound (VAMA) have been shown to be effective soil conditioners, though they are quite expensive and not simple to apply. With a proper knowledge of their incorporation into the soil, formation of good soil structure is possible. Although in the early 1970s they had virtually disappeared from the sales shelves because of their relatively high cost as soil conditioners, we may hear more about them at a later date because of their unusual structure-stabilizing property. In desert soils, these polyelectrolytes are best mixed into dry soil to a depth of three to six inches, and the soil wetted and allowed to dry to a condition where it may be tilled. After tilling and complete drying, the treated soil is ready for planting.

Wetting agents. Certain organic wetting agents and detergents have been shown to improve water infiltration rate and soil aeration, and to assist in the control of soil compaction. Surfactants reduce the compactability of cultivated peat soils. The beneficial effect of wetting agents to increase infiltration rate of water on highly water-repellent soils will vary depending on the dryness of the soil, the product used, and the texture of the soil. Before wetting agents are used it is best to consult with local state and federal agencies who have had experience with the various products in desert localities. Not all wetting agents work equally well on every soil and some wetting agents are not suited to desert soils.

Soils with serious compaction problems rarely respond sufficiently to soil conditioners to make it worthwhile treating them. Such mechanical tillage practices as deep tilling are required.

Group V

Soil inoculants (cultures). Except for the legume-nodulating bacteria (*Rhizobia*), general soil inoculants are valueless. Inoculation of seeds, plants, or soil with cultures of microorganisms is not practical for soil-conditioning as is sometimes claimed. Soil cultures do not improve water relations in soil, break up clods, or make the soil more crumbly. Yet they persist as sale products still hounding the markets. The addition of foreign microbial inoculants (bacterial, fungal, algal, amoebic, or streptomycete) to the soil usually results in their rapid disappearance. Desert soils are teeming with native microorganisms held in check by a lack of food, waiting to multiply rapidly when any available carbonaceous material is added. The introduction of alien organisms into such a new and hostile environment rarely leads to their establishment. The kind, nature, and activity of the soil microflora are controlled almost wholly by the nature of the substrate available for attack. The very scarcity of a species or specific group of organisms is evidence in itself that the habitat is unfavorable for its persistence. For example, the large number of intestinal tract organisms entering the soil from animal droppings soon disappear, and any effect the organisms may have produced is transient. Organisms indigenous to the soil are in a much better position to compete for the limited available food supply than newly introduced organisms.

Inoculants of every conceivable nature have been studied, tested, researched, and researched again for over 100 years, yet *only* the *Rhizobia* legume-seed inoculant remains on the positive side of the ledger as being of economic benefit to agricultural crop production. The Rhizobia bacteria infect through the new growing root hairs to induce nodules to form.

Soil inoculations, other than the legume cultures that nodulate a specific legume, are not recommended for use in desert soils.

Soils contain vitamin and vitaminlike substances. For example, biotin, thiamine, riboflavin, p-aminobenzoic acid, pyridoxine, and inositol as well as traces of many other metabolic cell constituents have been reported in soils. Research, however, has long since shown that additions of vitamins and other growth substances, when added to soil, rapidly disappear. When plant and animal cells enter the soil, the growth-promoting substances they contain decay rapidly along with other cell constituents. Further research reveals that growth-controlling substances (vitamins, etc.) in the soil are due to the presence of living organisms which contain them in their tissues. Soil organisms possess the capability of synthesizing vitamins and growth-stimulating compounds for their own use, just as do higher plants. Even trace quantities of amino acids, ever present in soil organic matter, originate from the synthesis and excretion of the natural, active microflora. Only a small percentage of the soil bacteria requires growth factors. Vitamin B-12, for example, is required by only 7.2 percent of the soil bacteria. These more fastidious organisms obtain their B-12 from others that commonly excrete it.

Enzymes fall into a situation similar to that of vitamins and growth-controlling substances. Extracellular enzymes are extremely limited. No way has been devised to raise or lower the enzyme activity of the soil in the absence of soil microorganisms. The introduction of cultures of microorganisms, extracts, or liquid substrates from organisms possibly changes the enzyme activity in the soil. Enzymes are rapidly inactivated in the soil. Even if sufficiently large quantities could be produced and added to soils, they would be destroyed immediately by biological and chemical factors, as well as inactivated by physical adsorption onto colloids. The only way to alter the enzyme activity in soil is to supply the natural flora with additional food material in amounts sufficient to increase their numbers. Trace quantities of organic substance have no practical effect on the supply of enzyme, vitamin, or growth-promoting compounds in soils since they are rapidly decomposed.

Group VI

Tillage. Soils can become compacted by improper tillage and heavy traffic. Soil compaction reduces pore space and decreases air and water movement. Overtilling may also cause compaction. Whether plowing to the same depth or spading the home garden to the same depth each year, compaction will develop at the point of greatest penetration. Thus it is wise to vary the depth of tillage, when possible, to avoid soil compaction.

Stratified textures in soils also affect water movement adversely. Regardless of the change from sand into clay or clay into sand, water will be impeded at the junction of the different textures as shown in Figure 7.4. In fact, layering of any material in the soil, even if it is straw, disrupts the uniform flow rate of water. Mixing of soil textures in order to bring about a reasonable homogeneity in particle size, and the mixing of straw by tillage, favor a more uniform rate of water and air movement in soils.

Amount of Amendment To Use

The amount of soil amendment to use, of course, depends upon the specific material in question, and this is extremely variable. However, if the choice of soil amendments is limited to composts, manures, wood wastes, sludges, peats, and physical inorganic minerals such as vermiculite, sand, perlite, etc., recommendations are not quite so difficult. Turf, for example, requires a minimum depth

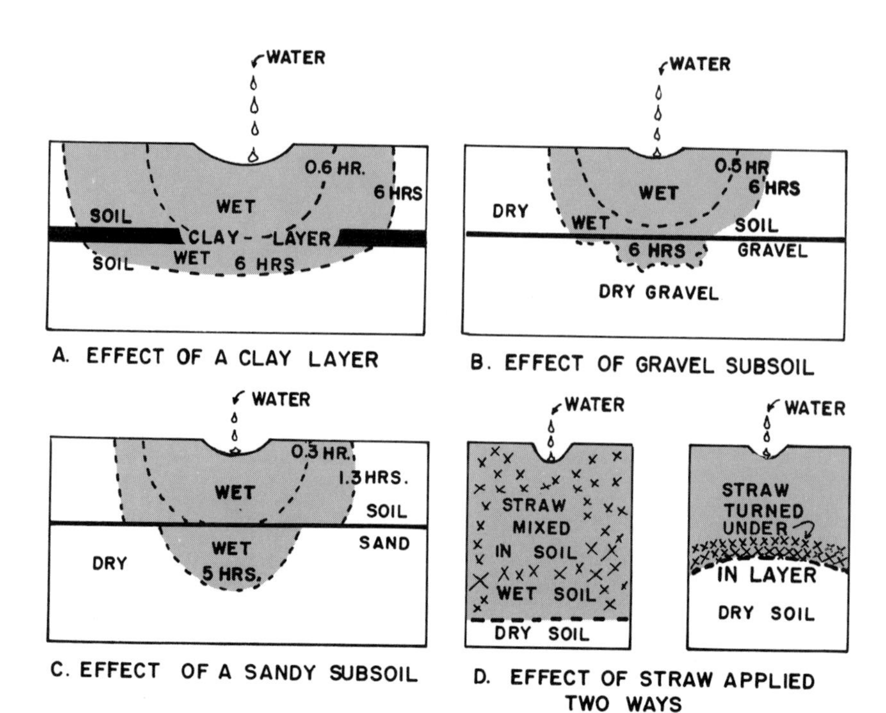

Fig. 7.4. Water movement in the soil is influenced by changes in texture and development of a non-homogeneous system. Thus, water and air move slowly across junctions between different textures whether they are mineral or organic (modified from W. H. Gardner. 1962. How water moves in the soil. Part 1, The basic concept. *Crops and Soils* 15:1–2).

of mixing. The upper six inches is optimum. Benefits from additions of five to ten percent by volume to the specified depths are questionable. The amounts should be at least 25 percent by volume. An average benefit range is 30 to 35 percent by volume. For example, an amendment to six inches depth and 30 percent by volume would require about 5.56 cubic yards of material per 1,000 square feet (Table 7.5).

Shrubs and trees grow best when the amendment is *mixed* into the soil throughout the depth of the planting hole. Where deep holes are used, additions of 15 to 20 percent of organic residue or manure by volume may be sufficient. Roses grow best when the soil mix contains 25 to 30 percent peat moss and salt-free, well-rotted manure.

Surface mulches around shrubs, trees, and in flower gardens should be at least one, and preferably two, inches deep. If feedlot manures are used, mulching with salty or fresh manure may burn tender plants. Deep soaking with water is recommended whenever manures are used on desert soils.

The amount of gypsum, sulfur, and sulfur-bearing compounds to use is given in Table 7.1 and discussed under the heading of Group I.

Earthworms

Earthworms abundantly inhabit moist organic soils where food is plentiful. Seldom do they appear in dry sands low in organic matter.

They are not detectable in the fine-textured desert soils when dry, but appear, as if from nowhere, immediately after a rain. Where they come from is considered by many a mystery. Homes established on virgin desert soon find an abundant earthworm population established in flower gardens. Experience indicates no need for seeding gardens with earthworms if moisture is adequate and a food source is available. Earthworms feed on fresh as well as decaying organic plant residues. Manured garden soils in Arizona have been estimated to contain as many as a million worms per acre. Unmanured soils may contain one-half or less of this number. Moist soils well-supplied with organic matter such as the Midwest prairie soils contain from 400 to 1,000 lbs of worms per acre.

Table 7.5

Soil Amendment Volume to Add for Various Depths of Treatment

Amendments as percent of amended soil	Depth of amended soil (Inches)						
	3	4	5	6	7	8	9
(%)	(Cubic yards per 1,000 square feet)						
5	0.46	0.61	0.77	0.93	1.08	1.23	1.39
10	0.93	1.23	1.54	1.85	1.16	2.47	2.78
15	1.39	1.85	2.32	2.78	3.24	3.70	4.17
20	1.85	2.47	3.09	3.71	4.32	4.94	5.55
25	2.32	3.08	3.86	4.63	5.40	6.17	6.95
30	2.78	3.70	4.64	5.56	6.48	7.41	8.33
35	3.24	4.32	5.40	6.48	7.57	8.64	9.72
40	3.70	4.94	6.18	7.41	8.64	9.88	11.13
45	4.17	5.55	6.95	8.33	9.72	11.10	12.52
50	4.63	6.17	7.72	9.26	10.80	12.34	13.88

Earthworm castings increase aeration, porosity, and water movement in soils. An earthworm will produce its own weight of castings in a day. Claims are made that as much as ten tons of soil, as castings, will pass through earthworms from one acre in one year. By their burrowing action they form channels for air, water, and roots to penetrate. Thus they aid in conditioning the soil.

Earthworms do not create fertility. Further, they produce no nitrogen or plant food. They grind organic residues, such as straw or roots, into fine particle size. Some of this is decomposed in the digestion process, and nutrients are extracted or removed by the worm. Their food requirements bring them to soils well supplied with organic residues. Reproduction is highest when residues are most plentiful.

The earthworm population under irrigated conditions in Arizona soils has been found to range from 10 to 350 per square foot. The average in good soils was 40, and in poor soils 10 to 20. The highest number ever reported in the United States, 350 per square foot, occurred in Arizona. These were found in a field with Papago peas chopped under and irrigated with an overhead sprinkler.

On grass ranges where the soil is wetted only by rains, no worms were found in the dry month of June. An average of 20 to 30 per square foot appeared in range soils after a rain.

8. How Arid Lands Are Managed

Some soils have been fouled so badly by poor management that they have had to be abandoned. In fact, one southwestern gardener claimed she couldn't grow flowers in her yard the way she used to and threatened to cover the area with green pea-gravel and "forget the whole thing." Unfortunately, this person is not alone. History records vast areas in world deserts which have been made worthless for planting by man's mistakes. Excessive grazing and tilling, and denuding land of natural vegetation, set destructive erosion forces in motion. Soil and sand erosion by wind movement continues to ravage large areas of once fertile land. Erosion of sand and soil moving on and off once stabilized land is serious enough in the desert, even under the best conditions of natural vegetation, without poor management adding to the problem.

Successful management must consider water as an integral part of the overall land plan. Again, historically, poor water management has allowed high water tables to form with no place for salts to escape. Such waterlogged conditions limit aeration and root growth.

The problems of management in the desert thus involve the necessity to integrate soils, waters, and plants in a planting program where: evapotranspiration is high; temperatures are high; soils are neutral to alkaline, and some are saline; soils compact readily; water of various qualities must be used for irrigation; estimation of amount and method of water application is critical to plant growth; only certain plants can tolerate the desert climate and soil condition; soluble salts are present in varying amounts and change within very short distances, even within a fraction of an inch; and leaching of salts out of the root zone is continually required to maintain a favorable salt balance, since salts are constantly brought in by irrigation water.

Soil Tilth

Management of soil and water centers around *tilth*. Tilth describes the soil's physical condition in relation to plant growth. An undisturbed as well as a cultivated soil may display good or bad tilth. A high-alkali soil which has developed under natural conditions is an example of one with inherently poor tilth, necessitating reclamation to make it productive. Excessive traffic over turfs and gardens, causing compaction, is an example of poor tilth where the owner lacks understanding of the fragile nature of desert soils. Soils in good tilth supply an adequate level of both air and water to plants. The pore space permits roots to extend to their maximum. Good structure and good tilth go hand in hand.

Extremes of soil texture (sand or clay) are associated with poor tilth. Sands are loose and have large pore spaces which drain excessively and are droughty. Clays tend to be cloddy and have small pore spaces which clog up and drain and aerate poorly.

Soil Compaction

Excessive traffic on soils when wet causes the structure to break down, particles to disperse, and puddling conditions to prevail. On drying, the soils become hard, crusty, and cloddy. Seeds germinate erratically, and those that sprout fail to emerge under the compacted conditions. Such soils exhibit poor tilth.

Along with the great advantages of modern farming and mechanization comes a serious problem of dispersed soil structure, compaction, tillage pans, pressure pans, and hard spots. Continued action of irrigation water compacts desert soils by weakening the structure. Pore space compresses, particles orient close together, and the soil becomes dense. Traffic of farm machinery, trucks, cars, people, and even animals over the land accentuates compaction problems. The pounding hooves of sheep, cattle, and horses can seriously compact range and other grazing lands. No soil is free from susceptibility to compaction, whether it is the irrigated farm, golf course, home lawn, flower garden, or our favorite football field. A good illustration of what is meant by compaction is seen by comparing two photos of cantaloupe root distribution in a compacted and noncompacted soil (Figs. 8.1 and 8.2).

Fig. 8.1. Vertical section of the root zone of cantaloupe in clay loam free from compaction. Roots are well-distributed in the soil (from Karl Harris, L. J. Erie, and W. H. Fuller. 1969. *Minimum tillage in the Southwest.* The Univ. Ariz. Agr. Expt. Sta. and Coop. Ext. Serv., Bul. A-39).

Fig. 8.2. Vertical section of the root zone of cantaloupes in clay loam having a compacted soil layer below the surface. Rooting stops at the compacted layer (from Karl Harris, L. J. Erie, and W. H. Fuller. 1969. *Minimum tillage in the Southwest.* The Univ. Ariz. Agr. Expt. Sta. and Coop. Ext. Serv., Bul. A-39).

Depending on their origin, there are three broad classes of soil compaction: (1) *geologic* — compaction formed in alluvial (water laid) materials when they were first deposited by water and glacial action, (2) *genetic* — compaction formed during the natural development of the soil, and (3) *induced* — compaction formed by mechanical pressure and vibration of farm implements and by the weight of water, man, and animals. Corrective methods are often the same for all three classes, although correction of compaction does not lend itself to generalizations.

A soil may be rather uniformly compacted from the surface to lower depths, or it may be compacted in layers. A compacted soil usually has an apparent density of about 50 percent or more than it has in its noncompacted state. It is not functional to assign a numerical level of density to define soil compaction. For the purpose of discussion, it is well for us to visualize a compacted soil as one whose apparent density is sufficiently high to adversely influence expected plant development. Compacted soils have some or all of the following characteristics: (1) a *pore space* that is greatly reduced, (2) a *water infiltration* rate that is slow, (3) a *soil aggregation* that is massive and less water-stable than usual, (4) an *air movement* that is slower and often lower in oxygen content than in soils of good tilth, (5) a *root penetration* that is limited in extent or completely blocked, and (6) a *root feeding* area that is drastically reduced and/or impenetrable.

Causes

Soil compaction has become a prominent deterrent in achieving maximum plant production with intensification of land use, such as new fertilizer practices, new insecticides, new crop varieties, irrigation, drainage, and other tools of greater plant growth. In the present stage of mechanization, the amount of traffic over land is exceedingly great. The weight and amount of this traffic over the home yard has increased with time and a greater desire to live outdoors. Many homes have heavy-duty machines which often are used without mercy to the soil. Several distinct types and degrees of compacted layers in soils result from this, depending on the implements used:

1. Compaction caused by relatively *light implements* used for seedbed preparation, such as the disc and harrow. These compact

soil at a shallow depth of about three to four inches, depending upon the soil type and mechanical setting.

2. Compaction caused by *heavy machinery*, such as tractor or truck. The surface few inches are often drastically compacted, water penetration is materially affected, and soil structures are broken into smaller aggregates or even pulverized into single-grain particles. Land planing of residential developments to level the soil requires heavy-duty machinery (Fig. 8.3).
3. Compaction caused by *plowing*, represented by plow sole compaction which occurs at the heel of the plow, is deeper than that of the other two types.
4. Compaction caused by *vibration* of machinery.
5. Compaction caused by the action of *irrigation water.* Soils seeded to perennial plants, such as turf, may develop hard spots as a result of compaction by water and reorientation of soil particles.
6. Compaction caused by the weight of *man and animals* on intensively used land. Much of this compaction is concentrated in the surface few inches.
7. Compaction and surface crusting resulting from *raindrop* (splash erosion) action on soils surfaces denuded and laid bare by man's removal of protecting vegetation (Fig. 8.4).

Fig. 8.3. A land plane leveling desert soil.

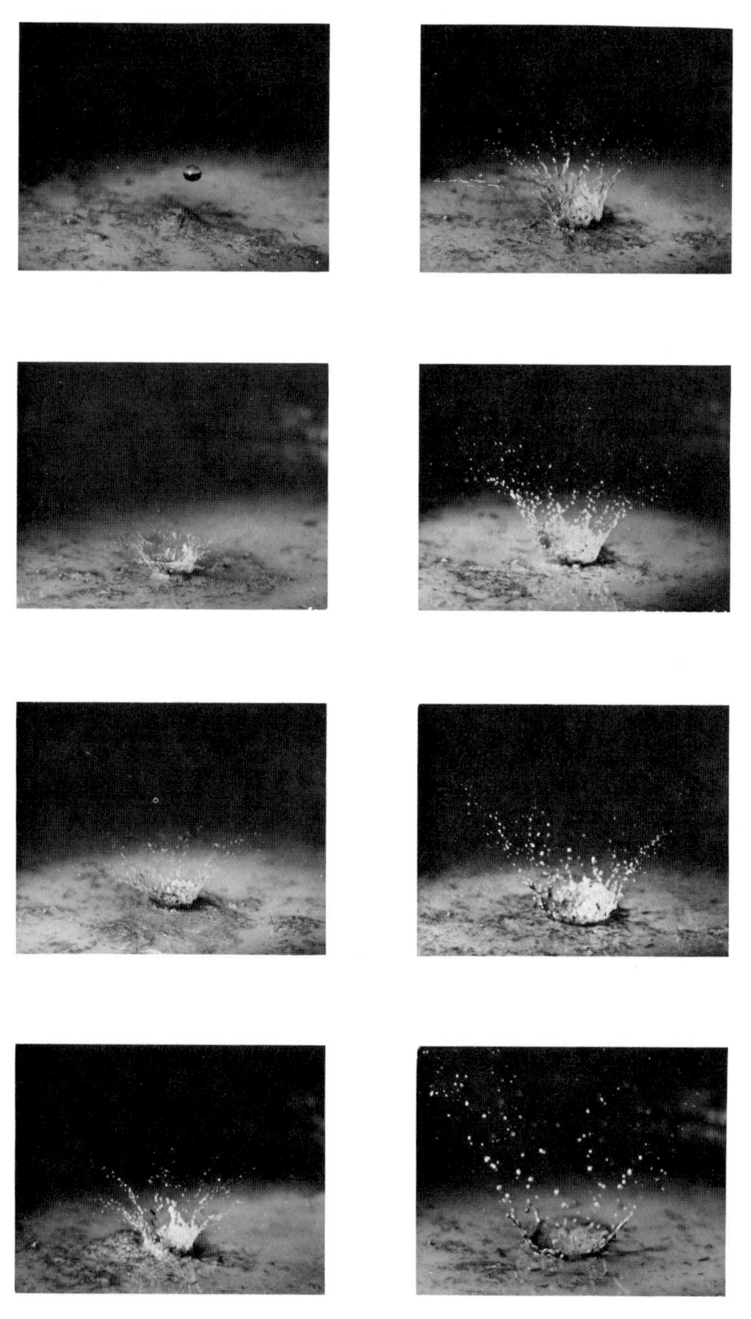

Fig. 8.4. The splashing action of the falling raindrop. Raindrops explode when they strike the surface of bare land. Each falling raindrop smashes into the soil like a bomb, scattering bits of shattered earth. Each drop is a baby Bikini explosion, and a rainstorm is a barrage, a bombardment of destruction. (Courtesy U. S. Soil Conservation Service)

Effects

Even the best growing methods used with irrigation cause a certain amount of soil compaction. The usual cultivation practices, such as equipment traffic, dispersion of soil structure during wetting, leave the soil in a condition that requires special tilling to prepare suitable seedbeds. The first signs of compaction may be noticed during preplant irrigation. *Water intake rate* is slow and irregular when a soil is compacted. For example, water penetrated a silty clay loam in the Yuma Valley at an average rate of 0.46 inches per hour during the first hour when rough tilled, but at only 0.09 inches per hour when excessively tilled.

The next critical period for establishing the extent of soil compaction is during the period of seedling emergence. *Crusting* of the surface soil is a form of compaction which often seriously prevents seed germination and seedling emergence. Poor plant stands and barren areas result from this form of compaction. Heavy rainfall and hand-sprinkling can cause crusting such that small-seeded plants do not emerge and reseeding becomes necessary. Lime acts as one of the cementing agents.

The presence of a plow sole or *pressure pan* as a result of tilling to a fixed depth, year after year, causes considerable deterioration of *plant growth*. Figure 8.5 diagrams what happened to plant growth on a silty clay loam when the pore space of the soil decreased to below 34 percent in a tillage pan three inches deep in the upper foot. Plants failed to become established.

Soils under perennial vegetation, such as golf turf and lawns, are highly susceptible to *compaction*. Paths, traffic lanes, and even traffic free areas may develop bare spots. The so-called *hard spots* in alfalfa, perennial pastures, and grass-seed turf which worry the farmer also appear in home lawns. The soil in hard spots is more compacted and takes water much more slowly than that of the surrounding area. Such spots require special soil and water management practices. Hard spots occur naturally when soils are placed under irrigation and in a vegetation pattern requiring only relatively infrequent tilling. They also occur as a result of the manipulation of land during home construction by soil fills of different textures being poorly placed. Particular recreation traffic patterns, irregular irrigation practices, and poorly placed organic materials (manure, peat moss, compost, etc.) in layers also cause hard spots. Hard spots take water so slowly that plants which become well-

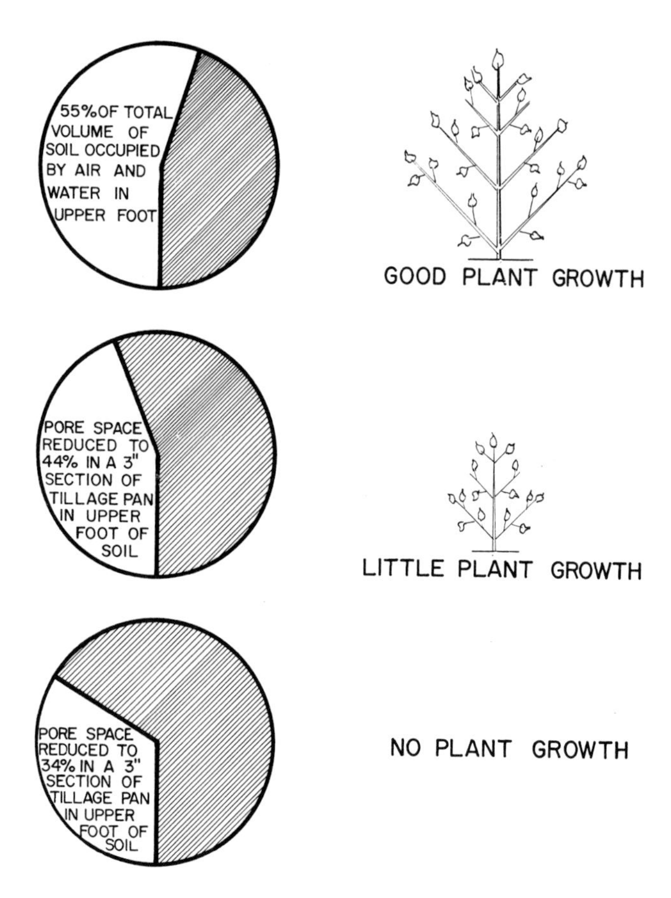

Fig. 8.5. Plant growth on a Mohave silty clay loam in the area of Mesa, Arizona, as influenced by a change in soil pore space.

established at first may actually die of drought during some succeeding year. These bare spots can enlarge each year until they occupy a sizable area.

Where land is sloping, compaction is conducive to *accelerated erosion* and *water runoff*. Various forms of soil compaction, therefore, may affect plant growth, limiting it in varying degrees from slight to severe, finally to where no growth occurs. In summary, soil compaction can affect plant growth in home plantings by limiting root feeding area through smaller pore space, seed germination, maximum emergence of seedlings, the amount of water available to plants, and aeration (roots require soil air to grow).

Correction and Prevention

Tillage practices. Some of the means of correcting or eliminating compaction through proper tillage practices are:

1. Reduce the traffic of heavy implements, as well as of men and animals, as much as possible.
2. Work the soil, tilling or spading, at the proper moisture content. Very wet soil compacts much more readily than dry soil. Tilling the soil when reasonably dry is the most important way of eliminating compacted soil. However, very dry soil should not be pulverized finely by excessive tilling or working with garden tools.
3. When preparing a seedbed, break up the soil through the compacted layers. In home gardens, deep spading when the soil is moist, not wet, will help break up compacted layers and make the soil more productive.
4. *Deep tilling* helps remove compacted layers as well as eliminate horizontal zonations caused by textural stratification during the formation of alluvial valleys and plains (Fig. 8.6). Deep plowing is recommended before placing a home on urban property.
5. *Knifing* with a horizontal blade is used to break up and lift compacted layers. It is less damaging to the soil structure than plowing.

Fig. 8.6. Deep plowing to break up stratification of soil textures and compaction.

6. *Ripping* is a process involving one or more tillage teeth set at various depths on a ripping machine to break up compacted layers.

Organic matter. As much of the plant residue as possible should be returned to the soil. Burning is justified only when the material is diseased. Organic residues should be worked into the soil as deeply as practical. Turning under organic matter to a depth of 6 to 12 inches is considered to be one of the most desirable methods to improve soil tilth.

Plant rotation. To prevent the formation of hardpans in the same spots each year, plantings should be rotated. Certain plants, because they are deep-rooted, are effective in improving water penetration and aeration on lands having heavy clay subsoils.

Traffic. During the planning of lawns and large gardens, convenient walks should be placed where people tend to make paths. Footpaths and dog traffic can seriously compact wet soil. When it becomes inevitable that paths will persist in established areas because of traffic, permanent pathways or sidewalks may be installed. Barriers of hardy plantings strategically placed may prove effective in creating a more desirable traffic pattern.

Minimum Tillage

Minimum tillage reduces the effects of soil compaction. Minimum tillage means that tillage practices are adopted not only to keep mechanical manipulation of the soil by man to a minimum, but that tillage operations necessary to prepare an acceptable seedbed are aimed at minimizing compaction. One form of minimum tillage is often referred to as *rough tillage* (Fig. 8.7) and includes (1) plowing or spading to a depth below the compacted layers and at a moisture level at which the soil will turn up cloddy, (2) irrigating, and then allowing the clods to dry somewhat before final tilling, and (3) using the least tillage necessary to prepare a seedbed.

Minimum tillage does not mean that the seedbed is left so cloddy that seeding is impossible. Irrigation of the soil when rough, just out of plowing or spading, has a favorable effect on soil tilth. The clods decrease in size and the soil surface becomes smoother, as if tilled. Moreover, water penetrates better to a depth where it can remain in storage for plants as they develop. Deep irrigation encourages roots to go deeper and extends their feeding area. The amount

Fig. 8.7. Rough tillage before irrigation (from Karl Harris, L. J. Erie, and W. H. Fuller. 1969. *Minimum tillage in the Southwest.* The Univ. Ariz. Agr. Expt. Sta. and Coop. Ext. Serv., Bul. A-39).

of tillage required will depend on the kind of soil, organic matter content, plant to be grown, method of irrigation, time of year, size of seed, salt concentration in the soil, soil moisture content, and weed control.

The principles of minimum tillage may be followed for either field or residential plantings. The less the soil is manipulated to attain the desired result, the better. Minimum tillage always includes the objective of first working the soil deep enough to break up compacted layers, hard spots, and areas where water infiltration is poor. If lime or gypsum layers are near the surface, deep tillage is to be avoided. These layers should not be brought to the surface and mixed with good soil.

Management for Control of Salts

The two principal sources of salts that may concentrate within the root zone to excess are irrigation water and the soil. Plants vary widely as to the amount of salts they will tolerate in the soil solution. The suitability of arid and semiarid soils for growing productive and/or beautiful plants depends on the manner in which soil, water, and plants are managed for salt control.

Salt from Water

All water contains some salt; therefore, continually adding water to the soil will add salts. The amount that will accumulate depends not only on the quantity of salts in the water, but also on the amount of water that passes through the soil. Such accumulation of salts can have an adverse effect on plant growth unless some means are provided to remove them from the root zone. Water is needed, consequently, not only to support plant growth, but to move salts downward *below* the root zone. Some common causes of salt accumulations resulting in the development of a saline or alkaline soil are frequent and light irrigation, poor soil drainage, high water table, insufficient leaching of tight, slowly permeable soil, and continued use of saline or excessively salty water.

The components of the water cycle and water sources available to a desert are generalized in a diagram (Fig. 8.8).

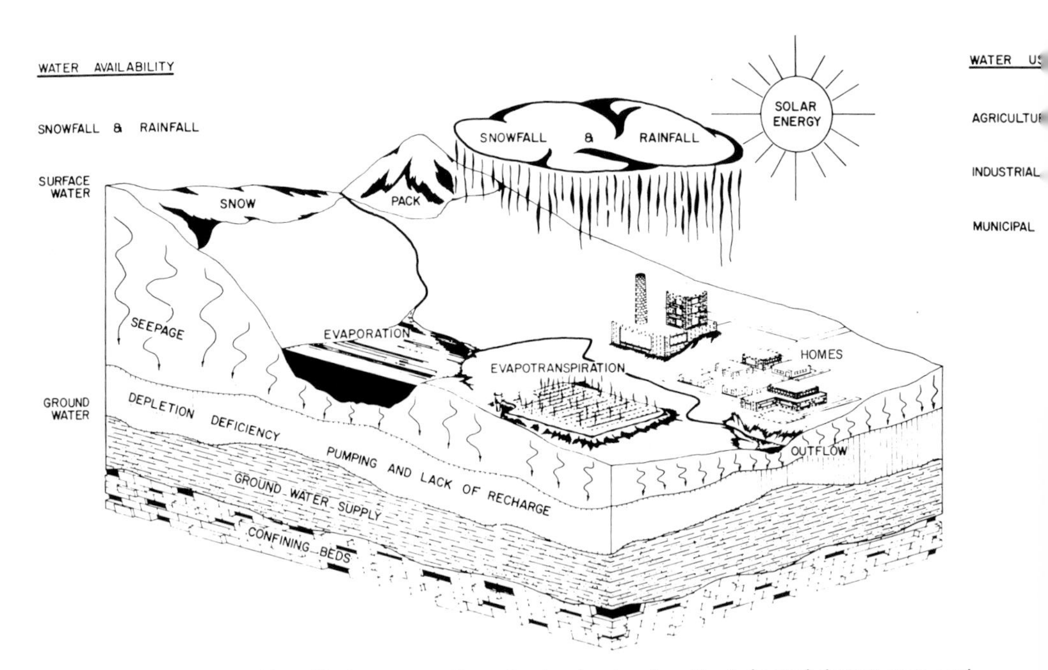

Fig. 8.8. Simplified cross-section of a basin showing the irrigated desert area and the Hydrologic Cycle that produces the available water (from W. H. Fuller, et al. 1968. *Efficiency of Water Utilization in Central Area of Arizona.* Ariz. Interstate Stream Commission Tech. Rep. No. 2).

Salt from Soil

Desert soils cannot be managed as indifferently as humid soils if salts are to be controlled. Improper management of soils will result in the movement of soluble salts upward by capillary action. The accumulation of salts on or near the surface of the soil as well as within the root feeding zone can reduce plants to poorer and poorer growth until eventually growth is not possible. There may be hidden zones of high salts below the surface, where compaction and textural changes occur. These layers must be broken up and leached to remove the harmful accumulations below the root feeding zone. Such layers or stratifications of accumulated salts can best be detected by sampling the soil every few inches of depth and testing for total soluble salt content. The concentration of salts throughout most desert soil profiles is not uniform. An example is shown in Table 8.1 for an alluvial soil from different locations.

How Salts Affect Plants

Successful landscaping and gardening in desert soils must also consider plant management practices. Plants differ greatly in their ability to grow in saline soils. Thus it is vitally important to know the specific plant characteristics with respect to salt tolerance level, placement of seed in the soil to avoid high soil-salt concentrations, rotation of plant species (or crops), germination and stand establishment, and effect of the osmotic tension of the solution bathing the roots.

Soluble salts in desert soils affect plants in various ways. Notably, they decrease the availability of water; alter nutrient availability; alter physical conditions of the soil, limiting root penetration; cause toxicity; and alter soil aeration.

The osmotic tension* of the soil solution is raised in direct proportion to the concentration of dissolved salt. In high concentrations, salts cause physiological scarcity (or nonavailability) of water in the root environment. True toxicity of common soil-salts to plants seldom occurs unless high concentrations of a specific toxic element,

*Osmotic tension — the pressure that influences the rate of diffusion (movement) of water through a semipermeable membrane (root cell walls).

Table 8.1

Some Examples of the Salt Characteristics of Soils From an Irrigated Valley in the Sonoran Desert Where Salty Water is Used for Irrigation

Depth of sampling (in)	Soil paste (pH)	Total dissolved solids* (mmhos**/cm^2)	Cations and anions in soil paste extract				
			Calcium (epm***)	Magnesium (epm)	Sodium (epm)	Chloride (epm)	Sulfate (epm)
0–12	7.9	7.2	30	12	53	42	51
12–24	8.0	6.0	34	16	44	28	49
24–50	8.0	7.4	31	12	60	34	57
0–12	8.1	7.3	29	14	60	30	58
12–24	8.1	10.2	37	20	78	69	54
24–50	8.1	5.5	28	12	38	17	50
0–12	8.0	5.9	26	12	45	20	54
12–24	8.0	10.0	27	21	89	56	67
24–36	8.1	10.8	26	26	96	61	68
36–50	8.2	8.5	26	18	80	35	75
0–12	7.9	21.3	62	41	142	177	45
12–24	7.9	13.2	35	27	110	100	61
24–46	7.9	11.5	34	26	98	73	69
46–50	7.9	10.3	29	22	90	80	53
0–12	7.8	20.0	60	32	130	169	38
12–24	7.9	16.0	39	27	144	144	53
24–36	7.9	19.0	39	38	155	158	57
36–50	8.0	13.2	36	22	122	109	63

*To get the approximate total dissolved solids in terms of parts per million (ppm) or milligrams per liter (mg/l), multiply by 700.
**mmhos is the abbreviation for millimhos, a measure of electric conductance.
***epm means equivalent parts per million.

such as boron, are present. The varying growth responses of different plants to saline soils, though, limit generalizations.

The most abundant salts in desert soils are those of calcium. Fortunately, they are only slightly soluble under the environmental conditions of most deserts. The abundance of sulfate and carbonate anions results in the dominance of gypsum and lime, both of which have relatively low solubilities. They give the surface of soils that white, powdery look. Problem soils also contain a prominent amount of the more soluble sodium and potassium carbonates, bicarbonates, sulfates, and chlorides. These salts give the soil a black, crusty appearance. They are highly soluble salts and must be leached below the root zone if suitable plant growth is to continue uninterrupted. A sample of soil sent to a soil-testing laboratory will quickly reveal the kind of salt which predominates in your yard or field.

Leaching Concept

Salts accumulating from water put on arable soil, and certain lesser amounts dissolved from the soil, must continually be removed from the root zone, or plant life will cease to exist. Water is the vehicle which can carry the salts away by a process called *leaching*. Leaching is necessary for desert soils regardless of what planting program is adopted. It consists primarily of adding excess water which washes out salts from the soil as it moves through the profile into the subsoil.

Leaching requires that more water be applied than is necessary to grow plants so that the excess will move *below* the root zone. This is the reason for recommending deep irrigation either at the end of the growing season or during the time of seedbed or plantbed preparation.

The amount of water needed to move salts below the root zone requires a knowledge of the quantity of water necessary to saturate a known volume of soil and the depth of root penetration of the specific plant growing. Table 8.2 gives a general wetting depth for soils of different textures. For example, it takes slightly more than one inch of water to wet a sandy soil to a depth of one foot. A little more moves the salts below the root feeding zone.

Water Management

All soils in the southwestern desert need to be irrigated to support successful home gardens. The amount and frequency vary

Table 8.2

Range in Available Moisture for Soils Having Different Textures or Ratios of Sand, Silt, and Clay

Texture range	Range of available water
	inches per foot of soil
Coarse Texture	
Very coarse sands	0.40–0.75
Coarse sands, fine sands, and loamy sands	0.75–1.25
Medium Texture	
Sandy loams and fine sandy loams	1.25–1.75
Very fine sandy loams, loams, and silt loams	1.50–2.30
Fine Texture	
Clay loams, silty clay loams, and sandy clay loams	1.75–2.50
Sandy clays, silty clays, and clays	1.60–2.50
Peats and mucks	2.00–3.00

according to plant species, specific climate (micro as well as macroclimate), time of year, and soil conditions. For example, shallow-rooted plants require more frequent watering with less water per irrigation than deep-rooted plants. Furthermore, less water is needed during cool than hot weather. Seedlings just getting established and transplants must be watered every day during warm weather if they are to survive and become well-established. Older plants may need a good soaking only once or twice a week.

The method of application of irrigation water varies. Each can be successful when managed correctly. The four methods are furrow, sprinkler, basin or flood, and trickle. *Furrow irrigation* is the most common method for row plants, narrow seedbeds, and low beds. Water is added between the rows and beds. *Sprinkler irrigation* is adapted to turfs. *Basin or flood irrigation* in yards and landscaping is confined to shrub and tree basins or used occasionally with small basin flower or vegetable beds. Commercially, basin or flood irrigation is used extensively for field crop production. *Trickle irrigation* was first used to an appreciable extent in the 1960s. Water is metered to standardized outlets where it continuously drips at a predetermined rate at a specific location such as near a tree. Large pecan orchards were being irrigated in this fashion during the late 1960s and early 1970s.

Time of watering. The most effective time of irrigation for leaching salts below the root zone where soils are only modestly salty is when the soil is driest. This usually falls at a time when the growing season is complete, or just before planting. In the desert Southwest, soils are driest in the fall and spring of the year. In the spring, preplant leaching as well as deep water storage is accomplished at the same time, even with a single irrigation.

A preplant irrigation to wet the soil to the depth of the expected effective root zone is good insurance for a successful garden. This means wetting to depths of two to four feet, depending on the plant to be grown. *A post-planting irrigation* is needed to wet the seeds and seedbed sufficiently to insure germination and seedling establishment. Surface mulching, with materials such as compost, well-rotted manure, bark, sawdust and straw, or even old burlap bags helps to prevent drying and crusting of the surface and holds moisture in the vicinity of the seeds and seedlings. As the plants continue to grow, the intervals between irrigation may be lengthened and length of time of wetting increased. The soil in the root zone must be kept moist continuously. Irrigation replenishes water transpired by plants and that lost by evaporation.

The nature of the soil, its location and texture, determine the amount of water it will hold and store. Sandy soils require more frequent irrigations than clay soils. Sunny south exposures also require more watering than shady north exposures, other things being equal.

Regardless of whether soils are used for home garden, residential or urban landscape planting, or farming, leaching of salts accumulated during the growing season should be undertaken at least once a year.

The time to irrigate plants is just before the plant needs water. This may or may not be the most desirable time for leaching. Signs of extensive wilting indicate irrigation is overdue. *More water* is required then than if irrigation is begun at a time *just before the plant shows water stress.* Stressing of plants for water, under saline conditions, to the point where visual signs are apparent before watering, results in a reduction of plant quality. Plants do not always fully recover.

The quality of water also will affect the time of application of water for leaching. With waters of poor quality, more frequent leaching will be required. Under these conditions it may be necessary to combine leaching with watering sometime during the active growing season. Of course, this is relatively simple to do by using a little more water than is needed for plant growth, so that enough will go below the root zone to carry the salts from the root absorption area.

Frequency of watering. The frequency of irrigation to control salts depends upon several factors: quality of water, quantity of water, soil texture, permeability of the soil, depth to water table, time of year, salt tolerance of the crop grown, economics, the kind of plant, and stage of plant growth.

Quality of water. Various methods have been proposed to provide a rough estimate of the amount of water to use for leaching to compensate for water quality that is less than good. We know that the poorer the quality of water, the greater is the leaching requirement. Soils irrigated with salty waters must be kept at a higher moisture content than those irrigated with good water, because salty water is less available to plants under moisture-stress conditions. Most city water has a modest amount of salt. Private wells vary widely in water quality.

Quality of water for plant use is evaluated as much on its sodium content and ratio of sodium to calcium plus magnesium as on its total salt content. Soils are more difficult to manage when irrigated with waters of wide Na/Ca+Mg* ratios than narrow ones. Irrigation water should contain some soluble salts (other than sodium), however, to be considered of high quality. Some salt in the water keeps the soil flocculated and better structured. Rainwater, from heavy precipitation, has been known to excessively wash the salts from the surface few inches of soils, causing them to deflocculate and "puddle." Such pure, salt-free water disperses soils of the desert Southwest so extensively that they form an impervious crust upon drying, often preventing even the largest seeds from emerging (see Fig. 8.9). Such deflocculated, dispersed soil may be returned to production by applying gypsum.

Quantity of water. Passing water *over* the surface even when salt accumulations are present does not remove salts from the soil. Surface salts move down into the soil rather than up into the flowing irrigation water. Irrigation water moving across the land picks up

$$^{*}\text{Na/Ca} + \text{Mg} = \frac{\text{Sodium}}{\text{Calcium} + \text{Magnesium}}$$

Fig. 8.9. Soil deflocculated as a result of heavy rainfall excessively washing salts from the surface.

very little if any salt. To accomplish salt removal, water must move down below the root zone.

A rough estimate of the quantity of water required to wet the root zone and to carry salts down can be ascertained by knowing the soil texture and how much water is necessary to saturate a certain depth of a given texture, as explained in the section on leaching requirement (page 133). Experimentation has shown that it takes about one inch of water depth to soak the soil to one foot if the soil is sandy. Therefore, if a planting has a root feeding depth of four feet and the soil is sandy, 4 × 1 inch = 4 inches of water will wet the root zone. An additional one inch may be applied to take care of the washing of salts below the root zone. Thus a total amount of water five inches deep is adequate for leaching, plant use, and storage. All five inches must be added at a single irrigation.

If the soils have other textures, more water is required (Table 8.2). This method of water-use estimation is difficult to apply in your home gardens and plantings unless deep basins are provided.

Another way to determine the depth of watering is to use a soil auger or probe. Before planting, a test can be made by digging down with a spade to the depth of water penetration after allowing the irrigation to continue at a certain rate for a given period of time.

To achieve good water management, the permeability of a soil must be known. The less permeable the soil, the longer the water must remain on the soil to penetrate to the desired depth.

Water tables can cause trouble if they are near the surface. Fortunately, there are few soils in the desert Southwest with serious perched water table problems, except in narrow valleys along rivers. Where water collects near the surface, drainage must be provided. Drainage below the effective root zone of the plants must be adequate for removal and discharge of soluble salts. Caliche or lime layers may perch water in any soil where these layers occur. Where a caliche layer is present within the root zone of plants to be grown, it should be either broken up by tillage, or, if it is where a tree or shrub is to grow, a hole should be punched through it to allow for drainage.

Salt movement. Salts do not move in dry soil. They move by dissolving in the water added to soil and move with the water in the direction of the saturated flow.

During the *wetting cycle* salts move by gravitational and capillary forces. The highest concentration of salts in leaching water is found at or near the *wetting front* (Fig. 8.10 illustrates this diagrammatically). Salts do not move appreciably beyond the depth of water penetration, as the diagram illustrates. Therefore, if a quantity of water penetrates the soil to one, two, three, or five feet, or any other depth, salts will accumulate at or near the deepest point of penetration. Because of this principle, the history of past irrigation practices can be readily detected by comparing the soluble salt concentrations throughout the soil profile.

During the *drying cycle,* salts may move upward with the lowering moisture gradient. In other words, when a water deficit occurs near the surface, water moves upward by capillary action and brings some salts along from lower depths to the surface.

Soil Management

Soils of the desert Southwest are not different from world desert soils with respect to their requirement of special soil management practices if they are to remain productive. People new to arid and semiarid lands must adopt new practices and give special attention to soil as well as water management practices if adverse accumulation of salts and undesirable physical structural conditions are to be avoided.

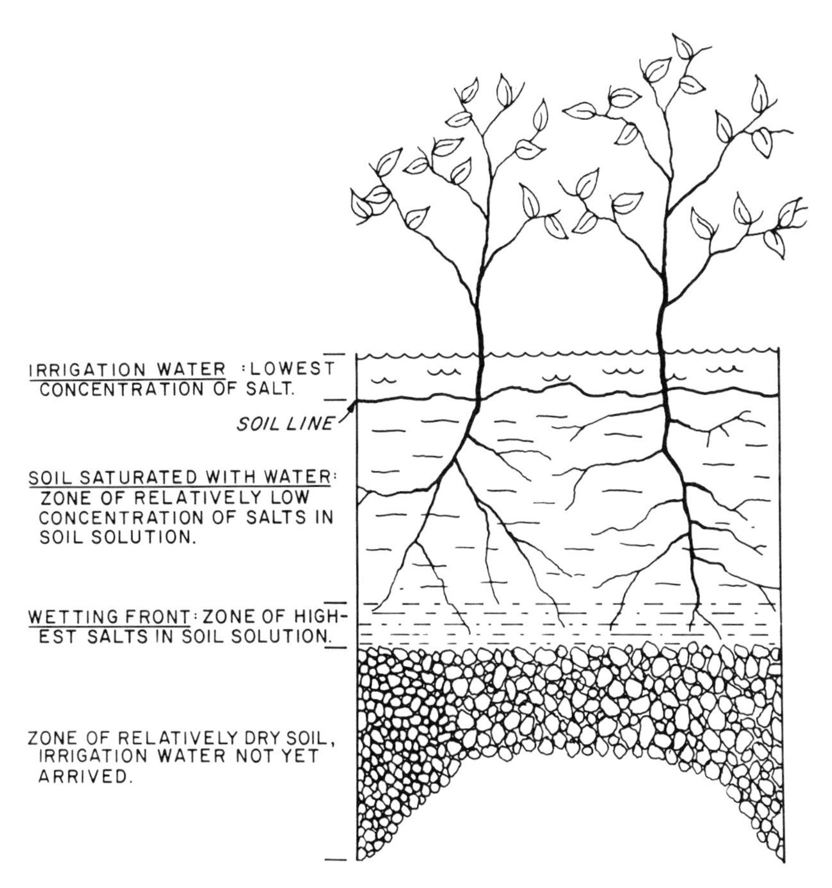

Fig. 8.10. Wetting of soil is illustrated ideally (from W. H. Fuller. 1967. *Soil, water, and crop management principles for the control of salts.* The Univ. Ariz. Agr. Expt. Sta. and Coop. Ext. Serv., Bul. A-43).

Tillage. Special tillage can do as much for the control of salts and maintain favorable physical condition of the soil as any other single practice. Deep cultivation can improve the tilth of hard, dense, baked-out textures, and clay and sand lenses. Minimum tillage, as discussed earlier, is part of this special practice to maintain good structure and a rate of favorable water penetration.

Leveling the land is a necessary step in the prevention of salt accumulation. The frequently observed half-moon dead spots on the upside of slopes is due to salt collections. Since slope is desirable for yard beauty, excessive sprinkling is often required to control salts. Accumulation of salts on high spots and ridges takes place where land leveling is irregular. Barren spots, often referred to as "hard spots" or "salt spots," may result from slight elevation changes imperceptible to the naked eye. Leveling often corrects poor plant growth associated with these micromounds where salts collect.

Plant beds are often deliberately ridged to allow for salt accumulation to pass up and beyond the seed row or the transplant. Figure 8.11 diagrams the positioning of seeds and plants to avoid areas of high salt concentration. Plantings should not be located on the high spots. The best placement is on the side or in the hollow of the bed. Salts in the soil solution thus move beyond the plant to the higher level where they concentrate harmlessly.

Organic matter and plant residues. Organic matter can play a vital part in the control of salts. As a part of soil management, no program should be long without the use of organic matter, even if it can extend only to returning crop residues to the land. Organic matter favors water penetration and movement of salts to lower depths in the soil. Large quantities of municipal wastes, primarily cellulosic in nature, could be used to solve both the depletion of soil organic matter and society's pollution problems.

Home owners can afford to use, in an advantageous manner, many organic materials and manures not economical to the farmer. Hard or salty spots can be corrected by working straw, compost, grass clippings, manure, chopped leaves or similar materials into the soil. These materials are most effective when mixed into a depth of 12 to 18 inches and watered to allow for soil improvement to take place.

Land selection. Within certain limits, land should be selected with a specific landscaping purpose in mind. All soils cannot be

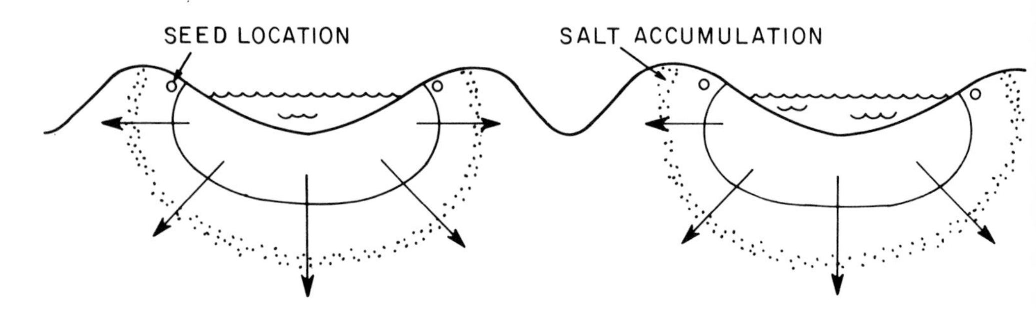

Fig. 8.11. Positioning of seeds or transplants relative to the distribution of salt concentration across a ridge in irrigated soil (modified from L. A. Richards, Ed. 1954. *Diagnosis and improvement of saline and alkali soils* (U. S. Dept. of Agriculture Handbook No. 60, Govt. Printing Office, Washington, D. C.).

managed alike. For example, more water must be used on silts and clays than on sands to maintain favorable salt balances. Some soils require deep leaching several times during the year and others only once.

Choice of soil. Landscaping of property around homesites often requires large quantities of soil for fill, lawns, playfields, and plantings. The quality of the soil moved into such areas is vitally important to the success of the landscaping. Generally red desert soil is recommended since the red color indicates the soils come from the better aerated upland, are freer of salt and lime than are gray soils along river banks and from areas of caliche accumulation. An independent chemical and mechanical analysis should be made before the soil is delivered. Sampling for analysis should reflect the representative nature of the soil. Sampling after delivery also is advised to evaluate the uniformity of the delivered material. Soil samples may be taken to the County Agent's or Farm Adviser's office for transmittal to the testing laboratory.

Plant Management

Plant selection. Plant management is an essential part of the overall control of salts. Selection of the proper plant in relation to salinity of the soil and water, seed placement to avoid loss of plants during the critical germination and seedling stage, stand establishment, rotation of plants, and an understanding of the relationship between salt of the soil solution and plant growth are important plant management principles to know if a permanent yard or landscaping is to be enjoyed.

Relative salt tolerance. Plants differ markedly in their tolerance to salt. Varieties within the same species also vary. In crop plants, for example, different varieties of cotton, barley, and grain sorghum have different abilities to grow and produce well in salty soils. Similar variation in salt tolerance characteristics is apparent in ornamentals. As an illustration, roses differ widely from pyracantha in their tolerance to salts. A table showing the relative tolerance of certain plants to salt has been prepared by the U.S. Department of Agriculture and is reproduced in part as Table 8.3. The values in the USDA table are relative, and the choice of suitable salt tolerant varieties should be evaluated with reference to the conditions under which each plant variety is grown.

Table 8.3

Relative Tolerance of Crop Plants to Salt*

Fruit Crops		
High salt tolerance	**Medium salt tolerance**	**Low salt tolerance**
Date palm	Pomegranate	Pear
	Fig	Apple
	Olive	Orange
	Grape	Grapefruit
	Cantaloupe	Prune
		Plum
		Almond
		Apricot
		Peach
		Strawberry
		Lemon
		Avocado

Vegetable Crops		
Approximately 6500–8000 ppm	**Approximately 2500–6500 ppm**	**Approximately 1900–2500 ppm**
Garden beets	Tomato	Radish
Kale	Broccoli	Celery
Asparagus	Cabbage	Green beans
Spinach	Bell pepper	
	Cauliflower	
	Lettuce	
	Sweet corn	
	Potato (white rose)	
	Carrot	
	Onion	
	Peas	
	Squash	
	Cucumber	
ECe $\times$ 10^3 = 10 to 12	**ECe $\times$ 10^3 = 4 to 10**	**ECe $\times$ 10^3 = 3 to 4**

*In each group, the plants first named are considered as being more tolerant and the last named more sensitive.

From L. A. Richards, Editor. 1954. *Diagnosis and improvement of saline and alkali soils,* p. 67, Agricultural Handbook No. 60, USDA, U.S. Government Printing Office, Washington, D.C.

Soil Reclamation

Perhaps the most serious problem facing inhabitants of arid and semiarid regions is the tendency for soils to accumulate salts to the extent of making them unproductive. Salt accumulation has often occurred naturally in the geological history of land. Man has accelerated it by irrigation. Abandoned farms are commonplace as a result of poor soil and water management and consequent salt accumulation. Also of serious proportions is the loss suffered from patches of saline and alkali soils, ranging from small spots to several acres. To the homeowner the problem results in ragged, chlorotic, and sad-looking plants where salt accumulation has been allowed to get out of control. Even homeowners need to know about reclamation procedures since there often are small patches of the garden or lawn which require special treatment.

Saline—Alkali Soils

Soils are divided into *saline* and *alkali* categories for convenience. A saline soil is a salty soil which technically has about 2800 ppm dissolved solids (salt) in the saturated soil solution. These soils are the ones which were referred to earlier as having a white, powdery surface. An alkali soil is sodium-affected. These sometimes are called "sodic" soils. These soils have black, crusty salt. When soils contain both a high salt content and sodium, they are referred to as *saline-alkali.*

Saline, alkali, and saline-alkali soils require reclaiming. The process of removing these salts from the effective root zone of plants, permitting the soil to produce again, is called *reclamation.* It is necessary to distinguish between the saline and alkali soils to know what treatment should be imposed.

Reclamation Methods

Reclamation involves a number of well-tested procedures, as follows:

1. *Drainage.* Lowering of the water table to prevent salts rising to the surface by capillarity. This must be done before reclamation can possibly take place.

2. *Leaching.* Leaching moves salts by water transport away from the root zone.
3. *Chemicals.* Chemicals (agricultural minerals) are added to soils to provide (directly or indirectly) for soluble calcium to replace the soil sodium (primarily for alkali soils).
4. *Tillage.* Salts collect where the soil density, texture, and structure change. Tillage is necessary to make the soil homogeneous and counteract compaction, improving the rate of water penetration.
5. *Crop management.* Certain crops are more effective in soil reclamation than others. Bermudagrass (and other grasses) and alfalfa exemplify plant types used to assist in soil reclamation.

Soil chemical amendments. Alkali or sodic soils require some source of soluble calcium to replace *sodium.* The most inexpensive source of soluble calcium, not toxic to plants, is gypsum. Calcium provides good structure, and more rapid permeability of water and infiltration. Excess sodium causes the soil to disperse, deflocculate, and "puddle." Crusts, clods, and compaction are associated with a sodium-dominated soil. *Sodic soils,* historically, have been referred to as "black alkali," because sodium solubilizes the dark organic matter along with other salts. Sulfur, as well as sulfuric acid, can be used in place of gypsum to furnish soluble calcium, as explained earlier. The soil must contain lime, however, to react with the acid to provide the soluble calcium. Nonsodic but *saline soils* which have an excess of salts collect as white crusts and powders on the surface. Historically, these are called "white alkali" soils. These need only leaching with water and need no chemical additive such as gypsum.

Tillage. Tillage can materially assist in the reclamation of soils. Deep tilling, breaking up the compacted soil layers and stratifications of lenses and layers of different texture, helps in improving water penetration and facilitates reclamation in soils.

Saline and alkali conditions of the soil are among the greatest problems to the home gardener. Shrubs and annuals having great beauty, under normal conditions, turn yellow in the presence of excess salt. Such plants detract from the home, and are best removed and the salt problem corrected, using reclamation techniques. Alkali conditions must be kept under control if successful landscaping is to be continued.

Soil and Water Conservation

In a desert climate where rain is scarce and on broad, level valleys, the original concepts of soil conservation, which relate to the prevention of the eroding away of vast amounts of fertile topsoil into streams and rivers swollen with days of heavy rain, appear a little ridiculous at first sight. After a short time of living in the desert, this first impression disappears. Rains in the desert fall in torrents, washing out deep gullies and moving rock and debris down arroyos in spectacular fashion (Fig. 8.12). Then, as the land dries out again, strong winds pick up dust, sand, weeds, and debris, often carrying them miles from their origin. Man and his cultivation practices accentuate these problems, requiring the development of new conservation practices. The responsibility of developing soil and water management practices which enable maximum utilization of natural resources over indefinite time for the greatest good of all people is a real challenge to research agencies in arid and semiarid lands. The knowledge derived from arid land research is put into action by such

Fig. 8.12. A deep gully eroded in relatively level desert land as a result of an unusually heavy summer rainstorm. The gully started from a cattle trail on overgrazed land and a soil-pipe stratification.

agencies as the U.S. Soil Conservation Service, Bureau of Reclamation, Bureau of Indian Affairs, the State Agricultural Experiment Station and Agricultural Extension Service, and agricultural industries, as well as by individual farmers, ranch operators, and homeowners.

Soil conservation touches all mankind. Many homes in the desert Southwest built in low places in old drainageways and along arroyos have been flooded with mud because soil conservation practices were not followed by the developers. In California, mud flows on slopes near Los Angeles are notorious.

Soil and water conservation on desert lands includes the development of the following principles and procedures: (1) control of soil erosion caused by water, (2) control of soil erosion caused by wind, (3) control of sediment in streams and rivers, (4) alteration and regulation of streamflow by land, cultural, vegetative, and structural development, (5) soil and water management of cultivated lands, (6) watershed management, (7) maintenance of soil tilth, and efficient crop productivity, (8) maintenance of water quality, (9) efficiency in water use, and (10) nutrition of food crops as affected by soil-water-plant interrelationships.

Soil and water conservation is carried out through proper land use practices. Before such practices can be put into action as recommended, basic knowledge of what the land is like, kind of soils involved, and erosion characteristics must be described, inventoried, and classified. Only then can a sound action program be offered and suitable land capabilities suggested. This involves the development of detailed soil surveys, soil mapping and classification, and land classification. Other land features than soils must be known, such as topography, water-runoff information, land capabilities, and the general layout of the individual residential or ranch area. Climate, rainfall data, and growing season characteristics must be taken into consideration. Soil survey maps are available at U.S. Soil Conservation Service offices and State Agricultural Experiment Station communication headquarters.

Soil Classification

Soils are differentiated into classes on a basis of many characteristics. The classification system resembles many other systems of classification of natural bodies, yet there is a fundamental difference which makes soil science difficult to understand. In botanical taxonomy, a tomato plant, for example, is distinctly different from another plant and it can be readily identified. Soils, on the other hand, form a continuum, blending into each other without distinct boundaries. An individual soil must be fixed by definition, since it is representative of a spectrum of like characteristics with no discrete boundary between it and another spectrum of like characteristics. A classification system, therefore, must provide for changes, some of which may be difficult to foresee, yet which will fit into the system. The soil is a complex system involving all basic sciences. It integrates biological and physical systems into a unique living body.

Soil survey. Based on a natural classification of soils, intended to encompass all lands of the world, a survey is conducted to inventory this most valuable resource at any given place. The inventory serves many useful purposes. To mention a few, it provides a basis for (1) evaluating a natural resource for planning present and future production of food, fiber, and shelter; (2) developing soil and water management practices for most effective and efficient use of a natural resource; (3) developing a tax base for resources evaluation and tax structure; (4) planning recreational facilities as best use of land; (5) planning roads, urban development, industries, zoning restrictions, and conservation of natural resources; (6) farm planning in soil and water conservation, land use and capabilities, crop use, stock ponds, fertilizer programs and many other agricultural programs; (7) planning residential use of land and general landscaping programs; and (8) obtaining bank loans and for real estate businesses.

These represent only some of the practical uses of soil classification, surveys, and mappings. Unfortunately, the great demand for soil surveys and interpretations for a multitude of uses cannot be met by the present survey staff associated with the National Cooperative Soil Survey of the United States Department of Agriculture.

Soils are delineated on an aerial photograph at a scale of about three inches to the mile, although some maps have a scale of one inch to the mile. The trend is to provide more detailed information. Each soil delineation is provided with symbols which give the *series* and other pertinent physical and morphological information. A series is a group of soils essentially uniform in their identifying characteristics and in arrangement of horizons. When identifiable distinct horizons are thin or absent, a series is a group of individual soils, the material of which is uniform in soil properties necessary to identify with the series.

Land Classification

The soil survey map is essential for planning land *uses*. The soil map requires explanation before it can be put to practical use. Explanations or interpretations may be made for various purposes so planners will know how to use the soil survey data. Two such interpretations are those for agriculture and recreation. The system of interpretations is called land capability classification. Other interpretations may be made for urban renewal, industrial building, flood control programs, roads, highways, ponds, lakes, public health installations, residential landscaping, and even home gardening. The kind of soil does make a difference in the successful establishment of an intended land use.

Land capability classification as developed by the USDA is available for general use. It relates to the way the soil surveys may be used for *land use planning*. This information can be obtained from the State Soil Conservation Service office. The two general divisions are based primarily on land either suited to or not suited to cultivation. Desirability of land for homesites takes other factors, such as "view," into consideration.

Erosion Problems

Soil resource problems are related to conditions prevailing in a desert environment. Some environmental conditions conducive to erosion problems are: (1) sparse or no vegetation, leaving soil bare to the climatic elements; (2) winds which sweep clean at accelerated rates; (3) dry surface soil much of the time; (4) loose sand, sand dunes, and blowout pockets; (5) infrequent but torrential rains: high velocity and intensity rain with little vegetation to intercept and reduce its fall; and (6) easily dispersed soil, often with a low degree of water-stable aggregates.

Erosion by wind. When dry and loose, all textures of soil are subject to movement by wind. Sand dunes may move at a startling pace and stop only when wetted by rain and when not moved by water action. Silt and clay blow most when very dry, particularly when they have been loosened by tillage instruments in seedbed preparation and until vegetation covers them. Sometimes granulated clay also will drift.

Sands abound in desert environments. They often appear to dominate soil textures, and indeed they do in most soils. Clay is the next most common texture, with less silt than is found in soils of humid climates. Surface sand can cause serious denudation of desert vegetation when it is blown across more stabilized soil surfaces. Damage to plants can occur by burying them and/or shearing off seedlings. Creeping sand dunes may bury well-developed soils and valuable home lands. Control of wind and its damaging effects in desert areas is a continuing subject of world study.

Shifting sand can have a marked effect on soils, even though it may cover the land only for a short period before it moves on again. Sands blown across desert areas may leave salts as a result of rains that wash and leach the salt carried on sand particles. This is thought to be an important mechanism influencing the salt content of desert soils. Gypsum dunes of sand-size particles are found in the desert Southwest. The White Sands of New Mexico are a good example of these. Lime dunes appear in Death Valley, California.

Sand dunes form many morphological structures across the landscape. These "squatter soils" occur as hummocks or dunes on rolling and undulating outwash plains, till plains, and flood plains, and as thin aeolian deposits over level terrain. Sands can develop profile differentiations in a relatively short time when stabilized under irrigated agriculture and urbanization.

Erosion by water. Whenever water collects on the surface and runs off, it carries some soil particles with it. The object of soil conservation under this circumstance is to provide conditions to reduce the erosion of soil to a minimum within the realm of economy of the land program involved. Even on relatively flat desert plains where infiltration is good, erosion by water takes place, though it may appear to be insignificant. When such land has been leveled and planted, even this natural erosion by water can be minimized.

Shallow soils with water infiltration barriers, such as clay layers, compaction, caliche, and other dense layers of material, erode seriously under heavy desert rainfall conditions, particularly where overgrazing has further denuded the land. As the proportion of bare soil increases, erosion further intensifies. Raindrop action has its maximum damaging effect on barren desert soils that characteristically disperse more readily than those from more humid climates. Sheet, rill, and gully erosion all occur in the desert.

Some of the factors influencing the extent of erosion by water are amount of rainfall, intensity of rainfall, duration of rainfall, amount and velocity of surface flow, characteristics of the soil, slope of the land surface, vegetative cover: density and kind, roughness of the land, rodent action in the soil, and traffic. Soil surfaces may gain as well as lose particles. Deposition of soil materials in low spots and valley areas can be very damaging to agriculture and other human uses of land.

Car and truck traffic on virgin desert land in the Southwest have been responsible for starting erosion that has ended as deep, scarred gullies. The scarring action of motorcycles, four-wheel drives, and other hill-climbing vehicles is often apparent (Fig. 8.13). Some of the very people who claim to champion conservation of our natural resources and wilderness areas have set into motion erosion processes with all manner of transportation vehicles, leaving the land scored with gullies, ruts, and washed-out roads. Ghosts of wagon trails still haunt the desert as eroded ruts, little healed even by a century of time.

Piping is one of the soil erosion problems peculiar to desert soils, including those of the Southwest. It can be a serious economic prob-

Fig. 8.13. Erosion caused by recreation vehicular hill-climbing on virgin land.

Fig. 8.14. Pipings in soil showing the outlet in an arroyo.

lem when man settles on the land and begins to concentrate water in the soil, either by irrigation or alteration of the natural drainage patterns.

Piping is a term used to indicate a tunneling erosion where the subsoil washes out from under the surface (Fig. 8.14). It has threatened the foundations of buildings, surfaces of school yards, highways, and parking lots. There are three types of piping: lateral flow of soil along a more stable subsoil into an open stream bank; flow along cracks by surface water; and flow taking place when dispersed surface soil is washed into pore spaces of coarse, poorly graded upper subsoil. Control of piping is accomplished by deep tillage, such as deep plowing and ripping or chiseling to mix the area of piping with soil from the surrounding area to obtain a uniform particle-size distribution. Small pipings have been controlled by wetting the land thoroughly and uniformly before tilling, cultivating, or using it for building purposes.

The control of erosion by wind requires as its first step the trapping of the soil and sand where it first starts to move. Control measures used extensively are windbreaks (trees, fences, shrubs, and other structures useful to the area); strip cropping where row or clean-tilled crops or plants dominate; rough tilling; dense plantings, cover crops, grass, shrubs, and any other stabilizing vegetation; and paving. Some of these measures call for community action. Any barrier or structure that tends to slow down wind velocity will decrease the load as well as the pick-up. Snow fences along highways can effectively keep sand from forming dunes on highways.

Control of erosion by water can be accomplished by initiating rather straightforward programs of: (1) altering topography to favor soil containment, (2) improving the rate of water infiltration through the favorable alteration of soil structure, organic matter content, fertility level, and soil-air volume composition, (3) increasing the plant cover by plantings, encouragement of natural vegetation, and agricultural crop rotations, (4) tillage, such as deep plowing, ripping, chiseling, mulching, contour planting, strip cropping, and diversion ditching and water spreading, and (5) straightening and deepening of rivers, lesser stream channels, and other waterways.

Rock Mulch

Historically, rock mulches were employed in arid lands of ancient Asia to conserve soil moisture for plant growth and control soil loss by erosion. A high proportion of world deserts are paved with stones and rocks naturally. Torrential rains and high velocity winds, so characteristic of arid lands, remove finer soil particles (sand, silt and clay), leaving coarser stones and rocks. Eventually the stones and rocks accumulate, blanketing the surface in either desert pavement or obstructing cobbles so well-known to the Gobi Desert. Soil erosion across these mulched areas becomes minimal. Thus, nature provides a lesson we may use where surfaces require maximum protection from erosion or where rock mulches practically and aesthetically fit into the home landscape design.

The high intensity of rainfall, though infrequent, coupled with sparse protection by vegetation and a natural tendency of the soils to disperse upon wetting, makes it necessary to protect all except the flattest of land. Rock mulches effectively and efficiently protect even highly sloping land. In one example, rocks were carried from an adjacent arroyo and placed randomly on the slope as close together as possible. Some weighed as much as 150 pounds and others very little. A variety in sizes, shapes, and composition added to the beauty of the mulched area.

The area represented by this example has remained fully stabilized for ten years despite the steep slope (15 percent), lack of vegetation, and erodible soil. During this time, rainfalls as intense as 2.5 + inches in one hour occurred two or more times a year. Yet neither soil nor water eroded into the swimming pool area below despite the soil's sandy clay loam texture. The six to eight inches of soil overlies caliche several feet thick.

Cactus is encouraged to grow among the rocks. Note the vigorous growth of the young saguaro among the rocks in Figure 8.15. Weeds are controlled either by a weedicide or pulled by hand.

Fig. 8.15. Native plants such as the young saguaro (*Cereus giganteus*) establish rapidly in rock mulches (photo by Clair Cameron).

Rock mulch has several advantages. It protects property from erosion; saves water otherwise required for turfs; makes beautiful, textured surface effects possible; requires little care — no mowing, watering, or fertilizing; permits collection of your own rock, which is good exercise and inexpensive; makes weed control relatively easy (paraquat or pre-emergence weedicide may be used); makes establishment of native plants easy; and permits economy of maintenance (no sprinkling system or periodic renovating needed).

Land Use Planning

Land use planning as a phrase has been around a long time. Even when the pioneer settlers moved across the United States, they recognized many uses of land, and that certain soils were better suited for certain purposes as compared with others: forest land, pasture and grazing land, and cropland, and special uses for urban areas, parks, highways, railroads, roads and lanes, and industry. Land was so plentiful, it wasn't until the early 1930s and the "Dust Bowl" days that we began to realize this heritage was fast deteriorating. The soils of our land were eroding to unproductiveness, and rivers and streams were filling with silt. The USDA Soil Erosion Service, renamed the Soil Conservation Service in 1935, began reversing this trend of land and soil deterioration. Land use planning came into practical being on a national scale through action of this agency.

Demands for special land and soil uses, mostly nonagricultural, increased sharply in the 1960s as the population bulge spread to the "end of the virgin West." Competing demand for the use of farmland for rapidly expanding urban and industrial developments and recreation expansions are extremely keen. Particularly during the late 1960s and early 1970s, level and fertile irrigated agricultural lands of the highest productivity have yielded to the explosion of urban building and specialty uses.

Land use planning no longer is limited primarily to agriculture. It is necessary to all interests of society. Land use planning, relating to the physical aspects, is largely concerned with soil and water management. Soils survey maps form the most desirable base of departure for land use planning, just as they did in the 1930s and 1940s for soil conservation, soil erosion control, and flood control problem solving.

Let us examine some (early 1970s) land use planning programs more recent than those developed for agriculture. Some good examples of land use planning for urban development may be found in central Arizona.

A city on developed virgin land. Northeast of Scottsdale, Fountain Gardens, a retirement city, has been developed on the footslopes of McDowell Mountain overlooking the beautiful Verde Valley. Land use planning is relatively easy under conditions where there are no seriously conflicting, prior land uses. Streets, sewers, flood control structures, utilities, and entrances and exits can be planned to fit into the topography, soil, and flood patterns with a minimum of conflict and expense. Legal and social problems are minimized. Land uses can be developed to take advantage of soil characteristics to minimize upkeep and management expense of landscape maintenance and provide broader choices of plant materials.

A city developed on level, fertile farmland. The McCormick Ranch on the outskirts of Scottsdale in the 1970s is developing into a city of 40,000 people. The soils survey map again furnished the basis for planning. Floodwaters of the main wash passing through the area are controlled by the establishment of lakes, parks, picnic areas, and golf links. The soil depth, texture, and structure had to be known before these and a water utilization program were developed. Caliche beds, rock and gravel strata had to be identified before lakes and other structures were excavated. Permeability and water infiltration rates of the soils in the general drainage watershed were required to plan lake sizes and numbers, and channel slopes through the new city. Clay soil was set aside for lake bed lining where lake displacement required digging into or through porous caliche and gravel beds. Thus, a serious flood hazard was changed into a beautiful landscape. Soil contours developed for golf course interest, home privacy, and aesthetics required detailed knowledge of the soil to depths of 6 to 18 feet to prevent bringing lime, caliche, and gravel unsuited for landscaping purposes to the surface.

An established city. The most difficult land use planning is that in an established city. In the case of Scottsdale, Arizona, as the city grows, more and more flood water accumulates on paved areas and streets which must find an outlet. Eventually the soil absorption and drainage channel capacities are exceeded and flood damage becomes severe. To accomplish flood control, removal of buildings and other structures was required along the drainageway of Indian Bend Wash as it passes through the highly urbanized area of Scottsdale. The soil in the alluvial drainageway had to be highly disturbed by necessary leveling, filling of gullies, and lake development for park and other recreational facilities. Again, good topsoil had to be set aside for turf establishment and clayey material stockpiled for lake lining. Thus, soil and water management practices played a major role in the land use planning.

Individual home. In contrast to *maxi* land use planning, there is what might be called *mini* land use planning. The latter is the planning around a single home. In the Southwest where water is so scarce and costly to the homeowner, land use planning is highly essential. Top priority for landscaping and gardening must take into consideration economical consumption of water. All of the soil management practices come into sharp focus in home landscaping. Placement of shrubs, trees, bulb gardens, and annual flowerbeds must be made with judicious attention to soil characteristics suitable for growing the plants under arid climatic conditions where water is such an important factor in successful landscaping.

9. Testing Soils, Waters, and Plants

The importance of fertilizer as a basic component for good plant growth is no longer questioned. What we should ask now is, "*How much* must be applied?" and, "*When* should it be applied?" The homeowner is demanding top performance from plantings around the yard, in terms of beautiful flowers, shrubs, and lawns, more than ever before. This demand on the soil is an essential part of the desire for a high standard of living through improvement of the environment.

When Justus von Liebig first advanced his mineral theory of plant nutrition in the 1840s, the problem of determining a plant's nutritional requirements appeared to be a simple matter of making a chemical analysis of the plant to determine its mineral content, and of the soil to learn its mineral content, and then comparing the two. It was soon learned that all of the nutrient elements found in soil are not available to plants. Elements like phosphorus, for example, exist in many forms, only a few of which can be absorbed by the plant root. There is enough phosphate in desert soils of the Southwest to provide all the plant requirements for several hundred years, yet certain plants are found to be deficient and respond to fertilizer application. A total analysis of the soil fails to reveal the availability of the constituents to plants.

Methods have been developed to measure *available* plant nutrients in soils. Chemical and physical analyses of the soil are simple and can be conducted by an analyst with only limited training. The knotty problem is interpretation of the analytical data, since the methods give empirical figures which, for the most part, have been correlated only broadly with responses of different plants on a few soils. Interpretation can be made only by someone with a thorough background of field experience in arid and semiarid climates. The County Agent or Farm Adviser is this someone. Factors other than nutrition may be responsible for poor plant growth. Chemical analysis cannot evaluate poor physical conditions of the soil, shallow caliche, droughty sand, poor watering, insect and disease problems, poor seed quality, or a host of other problems.

What Is a Soil Test?

The native fertility of our soil is being taxed to such an extent that the use of fertilizers is now a necessary practice where high quality and high density plants are desired. An awareness of the depletion of soil fertility, and the need for improved methods of soil and water management, have dictated a growing demand for a reliable means of measuring the fertility level. The *soil test* has been developed as a result of this demand. It provides a means of characterizing fertility and serves as a basis for predicting the requirement for fertilizers and soil amendments.

The soil test makes it possible to estimate the fertility status of soils. The test, however, supplies only part of the information necessary for predicting quantitative fertilizer or amendment need. A trustworthy interpretation requires a knowledge of history of the land with respect to fertilizer use, plants to be grown, soil management practices, quality of the irrigation water, and disease and insect problems.

Analyses by chemists in Agricultural Experiment Station laboratories are accurate and are correlated for desert soils over the western United States. Differences peculiar to certain soils are reflected in lack of ability of chemical extractants (i.e., those nutrients extracted from soil by chemical solutions) or soil solution always to reflect *availability* of the nutrient. This does not detract from the accuracy with which the chemist can quantitatively determine the amount of the element present in the extract or soil solution. The soil test, nevertheless, is useful for giving information where the available nutrient in question is excessive or seriously lacking in the soil.

A soil test can provide information only concerning the soil it represents. Therefore, a sample taken to a six-inch depth evaluates the top six inches of the root feeding area, which may extend to a depth of three to five feet. To be useful, the soil test must be related to growth responses on each soil and with different plants. When calibrations have been made relating a specified soil depth to fertility levels and fertilizer response, for example, then samples only of the first foot need to be taken. It is always good practice to dig down to at least three feet in some parts of the area to be sampled to determine the presence of caliche, compaction, stones, or other physical obstructions to roots.

A routine soil test evaluates the *nitrates* found in the soil-paste extract, *available phosphate, total dissolved solids*, i.e., soluble salts, and the *pH value* (acidity or alkalinity). Special chemical tests are made when there is some evidence that other *plant nutrients* are deficient or excessive. Mechanical analyses may be made for *gravel, sands, silt*, and *clay*, and the soil placed into a definite textural class. Analyses also may be made for the presence of toxic concentrations of certain elements, such as boron and fluoride.

How to Take a Sample

The most limiting factor in soil testing is *the representativeness of a soil sample for the area for which information is desired.* Each sample should represent a uniform area of like past history and, therefore, like composition. Separate samples should be taken for dissimilar and specific problem areas. Samples are easier to obtain when the soil is moist. Several separate cores, borings, or spade slices should be taken for a specified area, as shown in Figure 9.1. About

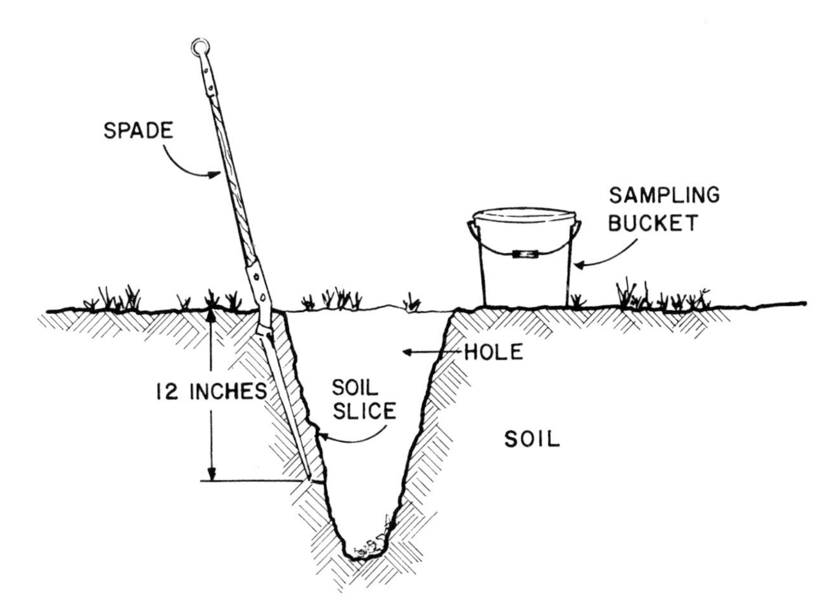

Fig. 9.1. How to take a soil sample.

one pint of soil is needed for routine tests. Around the home, a sample may represent a flower garden of only a few square feet, or a lawn, or a proposed tree hole. Regardless of size, only like areas should be combined into a master sample of several subsamples. The subsamples are first placed in separate buckets for each one-foot depth and broken up. Then they are mixed thoroughly. Finally, they are divided into one smaller sample per one-foot depth and packaged for taking to an independent laboratory or the County Agent's office for mailing to the state testing laboratory. Each sample should be identified as to owner name, location, date, intended soil use, sample number, tests wanted, and problem description.

What an Analysis Means

In desert soils, the *alkalinity* (high pH) and the *salt* content may limit normal plant growth. The soil test for arid and semiarid lands, therefore, routinely includes *pH values* and *total dissolved solids* (TDS) content. An excess of certain elements known to be toxic to plants is not unusual in desert soils and must be evaluated. The abundance of *potassium* makes any test for it unnecessary except for special plants and very sandy soils. Excess *sodium* (black alkali), which gives soils a poor physical structure, poor aeration, and slow water penetration, requires an analysis for *exchangeable sodium* and *gypsum requirement* to assist in outlining procedures for soil reclamation.

pH Value. High alkalinity, as indicated by a high pH value, can seriously retard or prohibit plant growth. One of the most critical pH values to look for is 8.4. Usually, soil contains excess sodium at and above this level, and reclamation procedures must be undertaken before planting. Reclamation is expensive and often time consuming. The majority of soils of the desert Southwest fall between pH values of 6.8 and 8.4. Only a few are truly sodic. Soils having pH values of less than 7.5 almost never contain alkaline earth carbonates (i.e., lime — calcium and magnesium principally).

Total dissolved solids. Total dissolved solids refers to the *salts* found solubilized in the extract of the water-saturated soil paste. Texture enters as a big factor in the interpretation of salt effects on plants. As a rough guide, salt concentration in terms of electrical conductivity of the saturated soil extract is given in Table 9.1 according to U.S. Department of Agriculture recommendations.

Table 9.1

Relationship Between Salinity of the Soil and Plant Tolerance*

Approximate total soluble salts	Plant tolerance
parts per million	
0–1300	Salinity effects mostly negligible.
1300–2500	Productivity and/or quality of very sensitive plants may be restricted.
2500–5000	Productivity and quality of many plants is restricted
5000–10,000	Only tolerant plants grow well and produce satisfactorily.
10,000 +	Only a few tolerant plants produce satisfactorily.

**From* C. H. Wadleigh. 1955. In *Water. The Yearbook of Agriculture*, pp. 358–61. Gov't Printing Office, Washington, D.C.

Nitrogen. Three forms of nitrogen are usually determined by chemical analysis in soil testing laboratories: *organic*, *nitrate*, and *ammonium* nitrogen. Laboratories receiving desert soils generally make tests for *total nitrogen* if a soil is found to contain an unusual amount of organic matter and nitrate nitrogen. *Nitrate* analyses always are included in the soil test where nitrogen fertilization has been heavy. No attempt will be made here to interpret specific values of soil nitrogen, since they can have little meaning without other necessary soil and plant information. An analysis of the irrigation water is helpful, also, in arriving at recommendations for reclamation.

Phosphorus. Desert soils usually are well-supplied with available phosphorus, although where intensive gardening is practiced deficiencies in certain plants may occur.

Potassium. Since plants of the Southwest only occasionally respond to potash, soil analysis is not made routinely for this nutrient. Soils should be analyzed when the land is first put into cultivation, and again after an extended period of planting or when soils are very sandy.

Mechanical analysis. Soil texture, i.e., the proportion of gravel, sand, silt, and clay, affects soil productivity. Soils excessively sandy are droughty and difficult to manage. Those excessively fine in texture, or high in clay, compact readily, take water slowly, and accumulate salts easily. Loams or sandy loams are often thought of as desirable textures for plant growth and ease of soil and water management. A textural relationship between sand, silt, and clay is given in Figure 9.2 to help in understanding relative size, number, and surface area per unit weight of the different separates.

What Is a Water Test?

The need for an evaluation of water quality in the desert Southwest was recognized even before the establishment of the State Agricultural Experiment Stations in 1889. However, it remained for the Experiment Stations to establish technical methods of analyses, and to evaluate water quality for agricultural purposes. Indeed, only in the last few years have the states been required by law to establish water standards for the control of quality of all surface waters. Fortunately, the quality of water from both surface and underground sources has been remarkably good in the Southwest, with only a few areas having brackish water. Yet by 1970, after years of withdrawals and recharging of lakes, reservoirs, and underground water supplies, water quality is deteriorating.

Fig. 9.2. Relationship between volume and particle size and number of sand, silt, and clay.

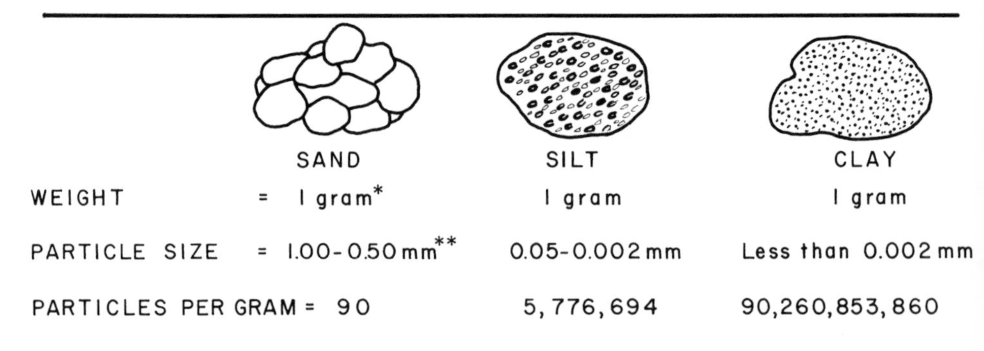

* one pound = 453.6 grams

** mm refers to millimeters which equals 0.03937 inch

The necessary use of large amounts of water under arid and semiarid climates requires careful attention to quality, which would be of little consequence under more humid climates. The high rate of water evaporation from soil and water surfaces, accompanied by a tendency for upward movement of salts through capillary rise in soils, the heavy demand by plants for water in a desert environment, and the poor rate of rainfall penetration for leaching salts downward, accentuates the accumulation of salts in the root zone of plants. Even the use of the prized water of the Colorado River, which had about 1.25 to 1.50 tons of dissolved solids or salts per acre foot in 1970, would add between three to eight tons of salt to each acre of soil each year, depending on the water requirement of the plant being grown. Home gardeners in some instances apply proportionately more water and consequently accumulate more salt per unit area.

Quality of the water plays an important part in successful plant production in desert soils. Salt-sensitive plants may grow quite well if the irrigation water is low in salt, yet fail to do well or even become established at all if the irrigation water is high in salts. On the other hand, salt-tolerant plants may produce quite well with water that would inhibit salt-sensitive plants.

Water quality is not only concerned with total quantity of salts, but also with the kinds of salts present. Certain elements, primarily *boron*, *lithium*, and *chlorine*, have a direct toxic effect at relatively low concentrations. Moreover, they may accumulate in soils sufficiently over a period of time to prohibit satisfactory plant growth, even though they may not be present in toxic concentrations in the water at any single irrigation. The *bicarbonate* ion also is considered to have specific ion toxicity, as is *sodium*, although the greatest hazard of sodium probably is indirect in its deleterious effect on the physical conditions of the soils.

What an Analysis Means

Water quality for irrigation has different criteria than does water quality for domestic use. Some factors determining water quality for plant use are (1) total concentration of *dissolved solids* or soluble salts, (2) relative proportion of *sodium* to *calcium* plus *magnesium* and, to a lesser extent, other ions, (3) concentration of elements such as *boron* that may be toxic in small quantities (the distribution of boron in Arizona waters is reported in Figure 9.3), (4) concentration of *bicarbonate ion* in relation to calcium plus magnesium, and (5) concentration of *lithium*, which is toxic in excesses of as little as 0.05 ppm with some plants which tend to accumulate this element in their leaves particularly.

Total dissolved solids. Waters having salts less than about 640 ppm give satisfactory growth on good soils. Waters with up to 1600 ppm can be used for irrigating salt-tolerant or desert-adapted plants. Waters having an excess of about 1600 ppm of soluble salts may be used only under certain conditions for irrigation with satisfactory results.

Sodium. Sodium hazard is evaluated by the SAR (sodium absorption ratio) according to a technical equation. An interpretation of the sodium hazard is made by the Soil Testing Laboratory chemist.

Boron. Sensitive plants can tolerate 0.33 to 1.25 ppm boron, semi-tolerant plants 0.67 to 2.5 ppm, and tolerant plants 1.00 to 3.75 ppm.

Bicarbonates. If the residual bicarbonate in water, calculated as sodium carbonate, has more than 2.5 ppm concentration, it is not suitable for irrigation; if it ranges between 1.25 and 2.50 ppm, it is marginal; and if it has less than 1.25, it probably is safe. Again these levels are guides only. Many factors in addition to the specific ion concentration influence what level is permissible for different plants.

Salinity and *alkalinity.* Irrigation waters have been classified by the interrelation between the salt and sodium concentration (for this classification see Fig. A.5, Appendix).

What Is a Plant Test?

If the chemical composition of the plant reflects the status of nutrients in the soil on which the plant grows, it seems logical that in analyzing the plant or a certain part of the plant the level of available essential plant nutrients in the soil should be reflected. Further, by knowing how much of each element is contained in a highly productive plant over the growing season, any deficiency in the soil should be exhibited by a negative deviation from productive plant

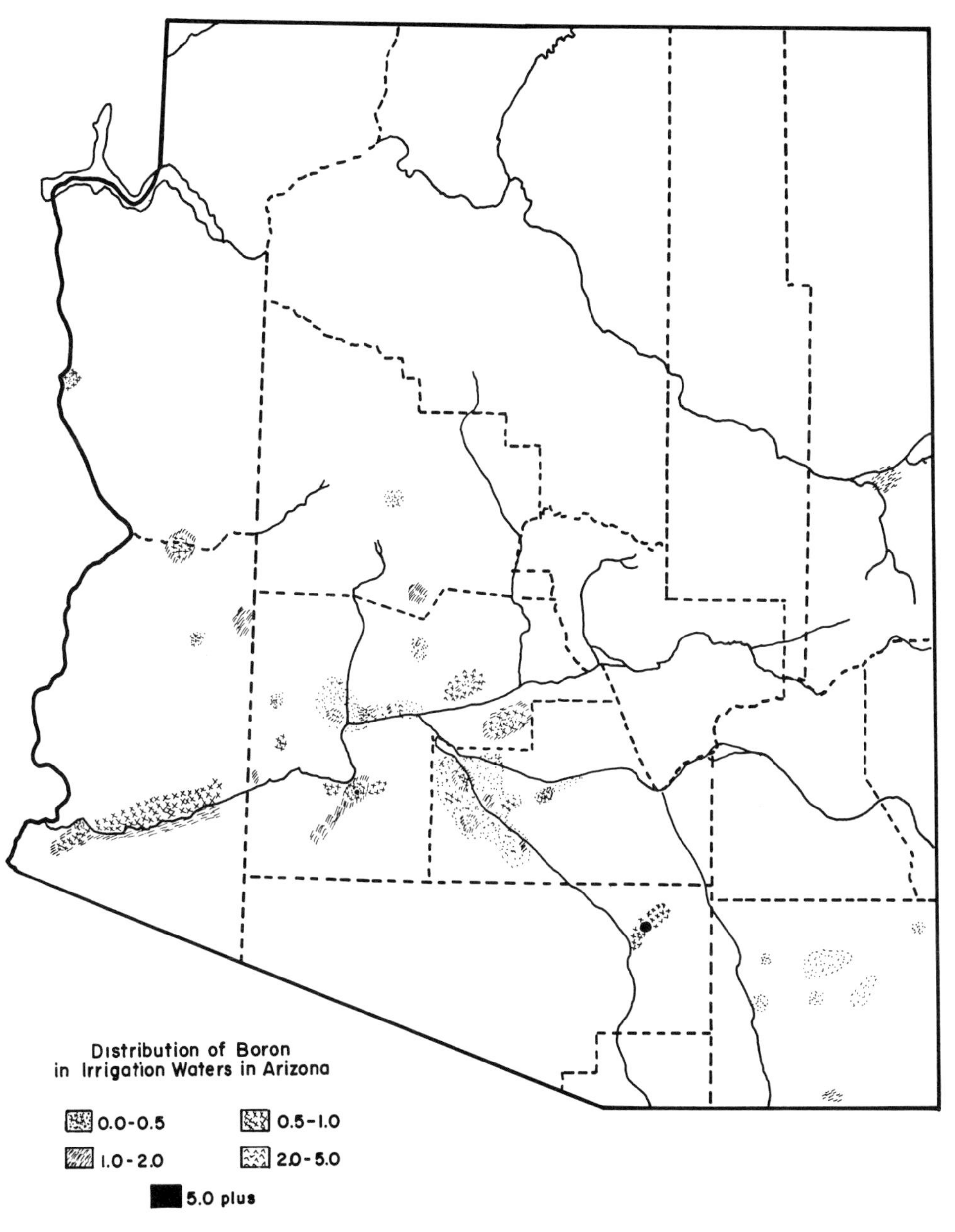

Fig. 9.3. Distribution of boron in irrigation waters in Arizona.

values. This interpretation forms the basis for plant analysis. Interpretation of the chemical data by experienced plant nutritionists is necessary.

Analysis of plant parts during the growing season is useful in pinpointing hidden deficiencies. It is a help to the grower in making fertilizer adjustments in season to help maximize production and plant quality.

What Tests Do Not Tell

Factors other than soil fertility may be responsible for poor fruit or vegetable growth and unsightly plant landscapes. Some of these factors are *biological* in nature: plant disease, insect damage, resistance to temperature extremes, poorly adapted varieties and species, and poor seed quality. Some factors are related to the *physical condition* of the soil: unfavorable soil-water relationships, soil compaction, surface crusting, hardpans and caliche layers, and shallow soil. Some factors are particularly important for determining the kind and location of plants: unfavorable location of plants, tree roots robbing smaller plants of water, caliche spots from excavated areas, traffic of man and vehicles, and poor seedbed preparation and time of planting.

These factors cannot be evaluated properly from a soil sample brought to the chemical laboratory in a container for testing. Evaluation of the conditions listed above requires on the spot inspection in addition to the chemical analysis.

Moreover, different plants have different nutritional requirements and have different abilities to absorb nutrients (feeding habits) from the different compounds found in soils. Vegetables require more fertilizer for maximum growth than cereal grains. For example, celery needs 1.6 times as much nitrogen, 3.3 times as much phosphate, and 7.8 times as much potash as does wheat.

10. How Soils Act for Waste Disposal

Man has been fighting the ill effects of his own pollution throughout history. Pestilence, vermin, plagues, and epidemics have stalked his life. Soils surrounding his early settlements became so impoverished that hunger and starvation set in and finally drove him to new land. Man moved continually to escape. Eventually there was little new land to which he, in his rapidly increasing numbers, could go.

The effects of the human population explosion on the environment, despite the warnings of a few in our society, went largely unheeded until after the middle of the twentieth century. Then mounting problems of many kinds forced the recognition that man could not increase his numbers at the same rate as in the past and still maintain the same high quality of living. One of the problems that awakened this realization was that of air, water, and soil pollution.

More than 90 percent of our nation's solid waste is directly deposited on or in the soil. Since air and streams are vehicles of transport, and human sensitivity to what enters the ocean is keen, only the soil is left for mass disposal. Therefore, management concepts assume a key role in our endeavor to use the soil intelligently as a *waste treatment system.*

The soil is an experienced "old-timer" at effective digestion and disposal of man's wastes (Fig. 10.1). Sanitary landfilling is widely utilized for disposal of municipal and industrial wastes. The application of engineering, biological, and soil science principles in waste disposal is required to insure a minimally detrimental, or even favorable, impact on the environment. With the wide variety of wastes to be disposed of, the increase in quantity, and the variations in methods of disposal, the pollution potential of leachates (percolating water) from disposal areas offers a hazard to the quality of surface and underground water sources as well as to the soil. The potential of erosion by wind and water resulting from transport of wastes across and over surfaces must also receive our attention.

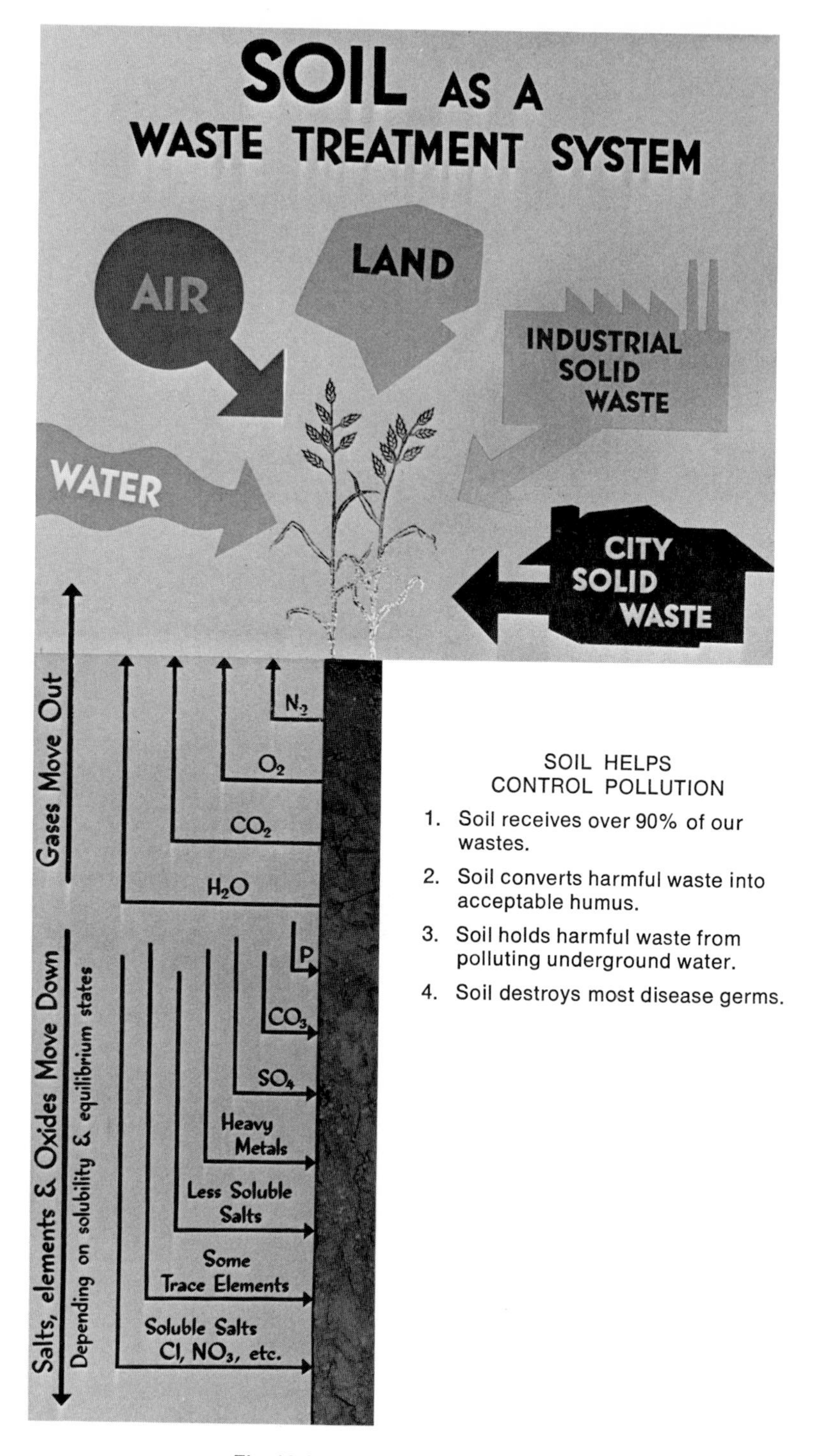

Fig. 10.1. Soil helps control pollution.

Soil as a Waste Disposal System

Fortunately, the soil is a versatile living system. Physical, chemical, and biological elements are continually reacting and interacting. The rate of interaction and dominance of one reaction over another is controlled by specific constituents in the soil. These constituents also continually change and vary with management. More specific factors — temperature, moisture, structure, texture, aeration, kind and quantity of clay mineral — and a myriad of variables interact to produce an unpredictable quantity and quality of leachate in any given soil site.

Basically, the soil is an *absorption* as well as a *biodegradation* system. Thus, both inorganic and organic constituents are altered along the depth of the soil profile and held from reaching groundwaters. Organic substances ultimately mineralize, yielding inorganic substances, gases, and water. The degradation (finally into such substances as water and carbon dioxide and small amounts of other gases — methane, nitrogen, oxides of nitrogen, sulfur, phosphorus — and into inorganic compounds, salts, and metallic oxides) reduces the harmful effect of man's wastes on his environment. Therefore, whatever is funneled to the soil will eventually pass out to the atmosphere as a harmless gas or into the soil as solids with minimum hazard, depending on the oxygen concentration in the soil and oxidized or reduced state of the constituents. The soil can be considered as a leavening agent, degrading different sources of waste to simple substances which are related to the original components of the earth from which they came.

Microbial Aspects

All material is subjected to microbial attack, though some inorganic substances such as glass and chromium, and organics like plastics, DDT, and oil, only slowly change chemical form. The soil is an ideal medium for organic residue decay. Wastes and debris are transformed by vast numbers of different kinds of microorganisms ever present and ready to act. In most soils the action is aerobic (in the presence of oxygen), although anaerobic (in absence of oxygen) conditions establish readily when oxygen is excluded by waterlogging or used up by overloading with highly decayable materials. Bacteria number into billions per gram of soil and fungi and streptomycetes into hundreds of thousands, regardless of the climate. The size and character of the soil population is controlled almost exclusively by the nature of the energy source (organic substances primarily).

Environmental changes may upset the natural equilibrium existing between species, kinds, and population densities, though the complex biota adjusts and readjusts readily to even the most severe habitat change. Even the desert soil surface is populated with a host of microorganisms, reflecting mostly the food conditions of the habitat. Fluctuations in population vary according to the energy source, just as the coals of a fire burn slowly, then rapidly burst into flame as new fuel is applied. For example, addition of carbonaceous wastes to the soil increases those organisms involved in carbon transformation. When metallic substances such as iron, copper, zinc, lead, and manganese appear in the soil, they will be oxidized or reduced depending on the aeration of the soil.

Despite the great volume of knowledge about soil microorganisms, homeowners with septic tanks and leaching fields still are confronted with a host of problems and little information on how to solve them. Studies on the movement of sewage effluent into soil under aerobic as well as anaerobic conditions show considerable slowing by clogging of the soil with microbial growth and slimes. To what extent this can be alleviated in soil under waterclogged conditions in the sides and bottom of disposal pits where limited air prevails was only beginning to be researched by the 1960s.

Chemical Aspects

Chemical reactions, not initiated or controlled by soil microorganisms, also take place. The soil's *air* composition is involved. *Carbon dioxide*, for example, unites with water to form an acid that reacts with the soil constituents as well as with materials put into the soil. The *oxygen* content affects the rate of decomposition or chemical change. The soil *clay minerals* control the exchange reactions of different elements and the tenacity with which elements are kept from moving into underground water resources. *Temperature and moisture* conditions influence rate of decomposition reactions continuously until the final products become permanently stored in the soil. Some reactions are slow, such as nitrate formation. *Solution* and *precipitation* reactions affect the ultimate deposition or movement of elements through the soil as they approach true equilibrium with the soil. Of course, the mineral composition of the water is influenced by the amounts and types of salts in the soil as well as those entering the soil as waste or from waste. All of the above factors control the elemental composition of the soil solution which may percolate into underground water sources.

Physical Aspects

The soil's physical characteristics, such as pore size, and distribution as related to structure and texture, are major factors controlling the velocity at which leachates move through the soil. Coarse sands and gravels allow more rapid movement of water than silts and clays, for example. Since pore size and distribution control soil moisture relationships, texture and structure also determine the concentration of the dissolved salts. These factors are highly heterogeneous in undisturbed soils of the desert Southwest.

Soil structure (architecture of the soil) can be altered by the kind of elements present in the percolating water and on its ion-exchange complex. For example, *sodium disperses* soils and *calcium flocculates* them. Salt concentration also influences pore size and distribution. Salts move through the soil by diffusion as well as in solution as leachate. Diffusion also must be considered when predicting the movement of specific elements (ions) in soils and their *attenuation* (retention).

Microorganisms in Pollution Treatment

Enormous quantities of debris become incorporated in the soil each year. Some substances degrade and change chemical form rapidly, others very slowly, some imperceptibly, but all eventually change.

Carbonaceous Substances

Carbonaceous residues in the form of above-ground as well as below-ground *plant parts* represent the largest single category of waste material. Literally billions and billions of tons cover the earth's surface. Animal and human residues, fowl, fish, and insect tissues add to this burden of decay, along with dead cells of microorganisms. If decay did not take place, life would long since have been smothered in its own debris (Fig. 10.2).

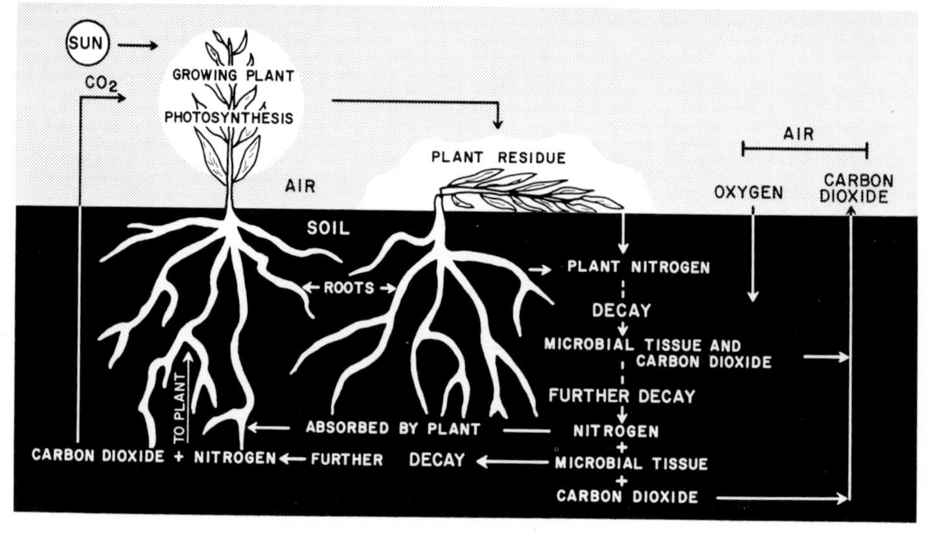

Fig. 10.2. Illustration of degradation of plant residues in the soil and fate of nitrogen and carbon.

The most abundant single organic constituent in nature is *cellulose*. Mature plants are approximately half cellulose, 10 to 30 percent hemi-cellulose, 5 to 30 percent lignin, and 5 to 20 percent water-soluble substances, nitrogen, minerals, and simple sugars. *Protein* constitutes a small fraction of the carbonaceous compounds. The rate of transformation of the different constituents varies greatly. Highly lignaceous materials (such as peat moss) decompose most slowly, whereas sugars and cellulose (grass and newspaper) decompose the most rapidly.

The rate of decay in the soil depends on moisture; temperature; ratio of carbon to nitrogen compounds; nature of the constituents (cellulose, lignin); aeration; and pH, salinity, and alkalinity.

Degradation of residues takes place more rapidly when they are mixed in the soil than when they lie on top. Plant and animal residues represent a large and effective reservoir of material for soil improvement. These sources need not pollute the land or end up in a garbage dump. They should be returned to the soil for its physical and nutritional benefit, either as compost or turned directly into the soil.

Municipal refuse or solid waste also decomposes more rapidly when incorporated into soil. When it is piled in a hole, a cell structure, or sanitary landfill, it breaks down slowly because oxygen is limited. Refuse (picked over to remove ferrous metals, glass, plastics, rubber, etc.) ground into fine material makes suitable *compost*. The value of raw municipal refuse (passed through a hammer mill and grinder, and screened) for improving the soil also has been demonstrated by agricultural crop tests. Certain nutrient and organic matter benefits were evident with no damage to the soil for plant growth when used in modest amounts. Thus one attractive solution to man's solid waste pollution is to utilize the waste for improving the productivity of the soil.

Carbonaceous wastes will yield *methane gas* (or natural gas) under the proper circumstances (Fig. 10.3).* Microbiological digestion or decomposition in the absence of oxygen produces about 72 percent methane, 25 percent other gases as carbon dioxide, ammonia, and hydrogen, plus a small quantity of mercaptans and amines. Between 50 and 80 percent of the organic material of municipal refuse yields gas. The remaining material is resistant waste and must be disposed of on the land. There always will be a residue remaining (primarily ash), of doubtful nutritional value as fertilizer. Disposal of this through the soil seems the best outlet. Methane from organic wastes is a renewable energy source as compared with fossil fuels. Moreover, methane is "clean." Enough gas to more than satisfy the current United States consumption could be produced from the present accumulation of solid waste.

Oxygen content of the soil air, in general, is quite similar to that of the atmosphere, but it fluctuates more. A large quantity of oxidizable substances like plant residues can be disposed of in the soil. They do not need to pollute an area. Easily decomposable leaves and grass or green plant residue do not need to go into the garbage collection to burden limited disposal facilities, but rather should be spaded into the soil. Wetting the soil enhances the rate of chemical as well as biological oxidation. We are all familiar with the rusting of metal equipment left in the rain. Constituents of all refuse change more rapidly into the more insoluble oxides under moist conditions in soils than in air. In the absence of oxygen, degradation and decay is relatively slow. *The soil is a natural and effective waste treatment system*, even under limited oxygen conditions.

Nitrogenous Substances

Nitrogen transformations are continually going on in the soil (as may be seen in the nitrogen cycle, Fig. A.2, Appendix). Nitrogen is one of the major plant constituents most susceptible to

*H. L. Bohn. 1971. "A Clean New Gas." *Environment* 13.

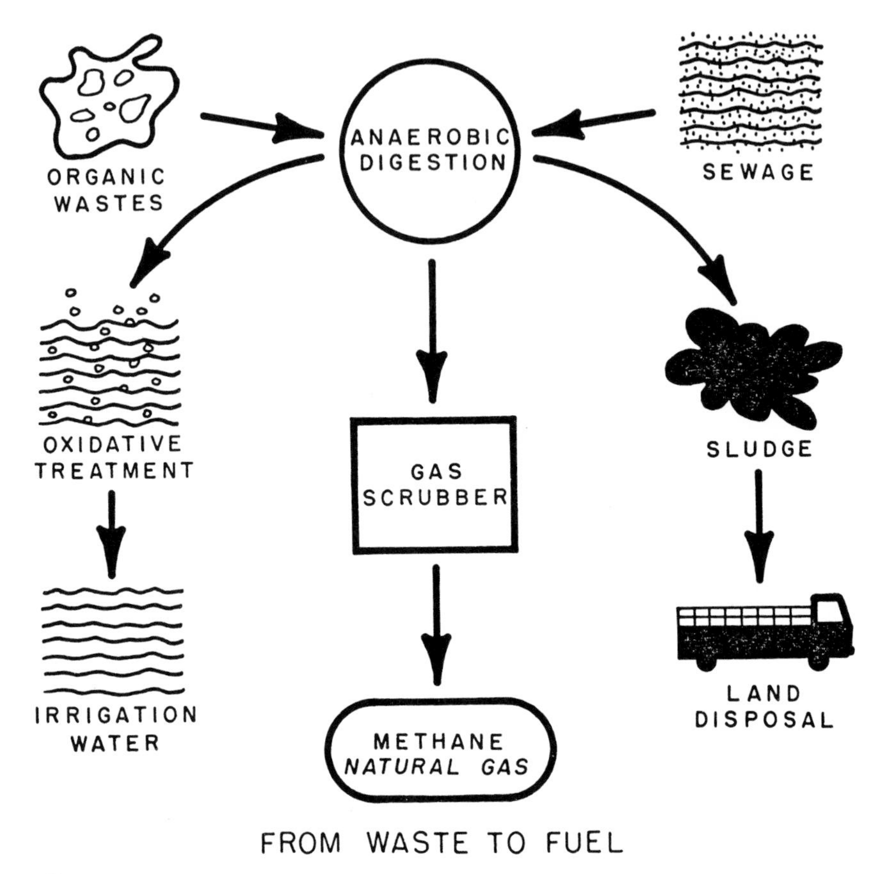

Fig. 10.3. A modified schematic diagram of the possible production of methane from organic wastes (from H. L. Bohn. 1971. A clean new gas. *Environment* 13 [10]).

biological actions. These include all changes from *organic* and *inorganic* to *volatile* nitrogen compounds. Loss of nitrogen to the atmosphere by several denitrification (decay of nitrogen compounds to harmless nitrogen gas, as N_2 in the air) pathways is a common and continual process and plays a role in pollution control. With the addition of easily decomposable organic materials to the soil, energy is available to denitrify and control nitrate pollution. Research has shown that very little if any fertilizer nitrogen reaches the deep groundwater of the desert Southwest. The natural nitrate content of groundwaters in deep wells in the desert Southwest may vary from a few ppm to over 400 ppm. This fossil nitrate is believed to have originated before man from plant and animal sources, being as old as coal.

Proteins are nature's nitrogen carriers. Protein, a constituent of all living cells, decays rapidly in the soil. Tissues of animals, birds, and insects are so fragile that their preservation, rather than their decay, is a problem. Waste from slaughterhouses, tanneries, and some meat and fish canneries offer problems of ready disposal and of odor. Decay "sets in" so rapidly that getting the refuse to a disposal site creates a problem. The soil is an effective neutralizer for these problems. Proteinaceous or nitrogenous substances, as blood and fish meal, make excellent fertilizing materials and are preserved for the market by drying. Processing cost for distribution on the market limits their most effective use for this purpose.

Phosphate Substances

Phosphorus exists in many different forms and readily changes roles. It is a constituent of meteorites, igneous rock, lakes, rivers, sea, and the soil. Phosphorus cycles readily between inorganic and organic systems. All life depends on phosphorus, since it is an essential constituent of living organisms and our very cells. (See the phosphorus cycle in Fig. A.3, Appendix.)

The early deposits of phosphates formed in prehistoric waters now represent the majority of phosphate-rich ores that are mined, as by the United States Tennessee Valley Authority Project, for example. Phosphates are being scattered to all parts of the earth, with man's relatively recent rapid development of uses for phosphorus in industrial and domestic chemicals.

As the human population continues to "explode," man's cyclic effect on concentration, utilization, and deposition of phosphorus will increase in geometric proportion to the population increase. Man is playing a critical part in the redistribution of phosphorus on the earth's surface. Some examples are:

1. Accelerating soil erosion with consequent loss of the organic soil layer, the layer of richest accumulation of phosphorus, to river bottoms and sea floors.
2. Redistribution of agricultural chemicals — fertilizers, pesticides, minerals — from phosphate ores to vast areas of the earth's surface poor in phosphate, where much of the phosphorus becomes fixed in the soil in a form unavailable to life processes.
3. Losses of crop phosphorus through sewage disposal that eventually "dead ends" in river and ocean bottoms.

4. Industrial and domestic losses through development of phosphorus chemicals, detergents, and solvents, and consequent disposal.
5. Pollution losses as gases or colloidal particles to the atmosphere, and final dilution by thin-layer distribution over the earth's surface.

Soil phosphorus contributes directly to redistribution in a reversible manner, as shown in Figure 10.4.

Sulfur-containing Substances

Desert soils, unlike those of many humid climates, contain sufficient sulfur compounds to provide all the necessary sulfur needs of plants. Sulfate (SO^{-}_4), the most highly oxidized form, occurs abundantly and is widespread. Gypsum ($CaSO_4 \cdot 2H_2O$), the calcium salt, is mined. Certain acid soils in humid climates require sulfur as an essential element of growth, but not calcareous soils.

Sulfur is also found in the organic fraction of soil, bonded to other elements. Soil microorganisms make this source available to plants through decomposition and mineralization of organic substances. The transformations of sulfur resemble those of nitrogen, the difference being primarily in the lesser solubility of the different compounds in the soil solution. Sulfates move much less readily in the soil than nitrates. Here again, the soil microorganisms play the key and vital role in making a nutritive element available to plants.

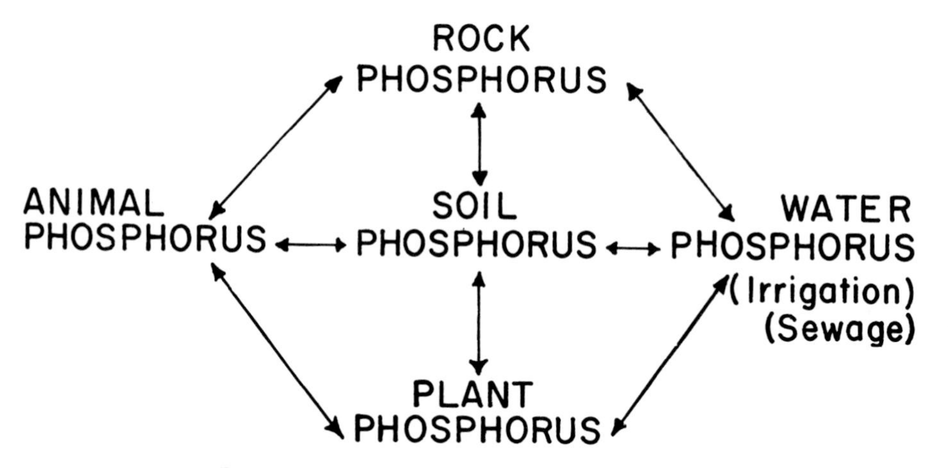

Fig. 10.4. Reversibility of soil phosphorus distribution.

The sulfur cycle in soils (Fig. A.4, Appendix) is dominated by certain major processes of (1) oxidation of elemental sulfur to sulfates and inorganic compounds as sulfides, polysulfides, thiosulfates and polythionates; (2) reduction of sulfates and other sulfur compounds; (3) assimilation and immobilization of sulfur compounds by incorporation into microbial cells; and (4) dissimilation or decomposition of organic sulfur compounds and complexes, with ultimate cleavage into smaller molecules and release to inorganic forms.

Desert soils differ from soils of more humid climates by having a greater proportion of sulfur in the inorganic than in the organic form. Gypsum is the primary sulfur carrier.

One must not confuse the effects of elemental sulfur and sulfate in soils. Elemental sulfur is the reduced form and sulfate the oxidized form. Sulfur must be changed into a soluble form (as sulfate) to be available to plants. Oxidation of sulfur to sulfate is brought about by soil microorganisms such as the bacterium *Thiobacillus thiooxidans*. A host of organisms are capable of oxidizing sulfur, and desert soils appear to be rich in sulfur-oxidizing organisms.

In Arizona more than 1,500,000 tons of sulfur (in sulfur-containing compounds) are dissipated annually into the atmosphere from smelter-stack gas. Oxidizing the gases to sulfuric acid to avoid air pollution allegedly would produce 4.5 to 5.0 million tons of acid, more than the agricultural industry can use for soil conditioning. Land disposal or a trade-off with a sulfur industry is suggested.

Micronutrients

Iron, copper, zinc, lead, manganese, cobalt, and molybdenum are subjected to microbial transformations in soil. They may be oxidized or reduced readily, depending on the oxygen concentration in the soil. Most of us are acquainted with the rapid oxidation of iron and tin cans in soil and the etching of brass (copper and zinc alloy) buckles. The subject is a book in itself. Thus, disposal of solid wastes of all kinds is greatly affected by the microbiology of the soil.

Management in a Soil Waste Treatment System

Soils function in many roles. The highest priority is in the production of food and fiber. The quality of the soil must be preserved to support this function as well as, if not better than, in the past.

The soil as a waste treatment system is another highly important function, which must include the aim of protecting and conserving the soil system in its support of life. Good management and intelligent land use are necessary to preserve these functions and others, such as foundations for buildings, roads, and recreation, to name a few. Soil management programs aimed at effective waste treatment must be executed also without adverse effects on water and air quality. Improper management or land use can end in pollution of soil, air, and water to the extent that they become a hazard for all life. It is to these multiple functions of the soil that management is addressed here.

Soil Pollution

Despite the great capacity of the soil to accept large quantities of wastes without interfering with its productivity, there is a point beyond which recovery is not possible. Excessive application of manure and compost is an example. In one case, the addition of 150 tons per acre of municipal compost to a soil, followed immediately with seeding to lettuce, reduced lettuce production to below that of untreated soil. Continued yearly applications at that rate reduced yields to zero. Moreover, certain toxic substances in modest amounts can irrecoverably pollute the soil. For example, certain toxic heavy metals are contained in sewage sludges and animal wastes. Such residues could conceivably pollute the soil from the accumulation of repeated additions. They require special disposal methods through the soil for their containment.

In the desert Southwest the soil can be polluted naturally, unlike in any other climatic region. Excess accumulation of salts, sodium, boron, or lithium takes place during soil formation, for example. Reclamation procedures are needed to recover the productivity of the soil. Growing crops tolerant to these elements is another solution to the problem.

It is not difficult to pollute the soil by addition of excess fertilizers. Excess application of phosphates on some lawns year after year ends with a soil problem of micronutrient deficiency, particularly iron. Grass quality deteriorates and the grass turns yellow and often dies. Excessive sulfur (yellow elemental powder) has been known to kill newly planted shrubs, whereas correct management of sulfur and fertilizer enhances soil productivity and plant quality.

Plant herbicides and soil fumigants have been known to adversely affect plant stand when the sensitivity of the plant seed was ignored. Sometimes weedicides remain active in the soil a long time and sensitive crops, even if planted the following season, may be adversely affected. Soil and herbicide management practices aim at avoiding such problems. Management aimed at providing the most favorable conditions for soil microorganisms to degrade and inactivate the many pesticides is a specific example of the way soils can function in pollution control.

At one time, radioactive fallout posed a threat to food production in soils, not only those of the desert Southwest, but all over the world. Radioactive strontium, a fission product of uranium, showed an increase in agricultural soils at the time of atmospheric bomb testing. Since plants cannot distinguish between radioactive strontium and soil calcium, methods of soil management were developed to eliminate or minimize the absorption of strontium into the food chain. Removing, or inactivating, the soil by various tested procedures, planting away from ridges, absorbing the radioactivity by some fast growing crop or plant cover and discarding the residue, are only a few control methods suggested. Ban on air explosions of test bombs and air missiles reduced this hazard at the source, following which the levels of radioactivity in desert soils reduced to the natural amounts. Since fallout correlates highly with amount of rainfall, the desert Southwest can expect a minimum contamination. Even during the most active period of nuclear explosions, the desert Southwest was relatively unaffected, having one of the lowest rates of fallout accumulation anywhere in the world.

In the pursuit of food and fiber production, and in the demand by society for a higher standard of living, soil erosion processes began accelerating as virgin land was broken by the plow. Soil management practices since the late 1930s have minimized erosion hazards. Strip farming, diversion ditching, crop stubble or residue management, mulch planting, grassing of waterways, and many other soil, crop, and water management procedures are being used effectively.

Management of soils of wild lands and overgrazed and stripped-off forests above our major dams still needs attention in preventing water pollution and sediment reduction. Erosion control of uncultivated range and forest land has yet to achieve the sophistication of erosion control methods of tilled and pasture lands.

Maintaining Soil Productivity

Long-lived wastes that may be absorbed and concentrated by edible plants should be placed where they will not get into the food chain and where they will not contribute to contamination of water. Though wastes have been put onto the land since life began and the decay of organic plant and animal residues is a normal and essential process, the characteristics of man-made wastes and chemicals are changing. The very slowly biodegradable plastic is an example. Industries create an increasing variety of inorganic as well as organic waste substances. Some of these substances produce an adverse effect on the soil. Trace and heavy elements or metals can adversely affect the physical structure of soil and inhibit adequate water movement. The quality of the soil for plant growth and even as a waste disposal system is impaired.

Economics of food production and processing which change with the times now dictate a high rate of waste disposal at or near the center of production. The cattle feed lot and food processing canneries are examples. Management of these wastes is needed to prevent overloading the soil to the extent of putting it out of production. Such management in the desert Southwest is considerably different from that in the Midwest, however.

Effluent from the Tucson, Arizona, sewage treatment plant is managed as irrigation water and applied to nonedible crops. Thus, water is conserved in a water deficient area and the effluent disposed of in a favorable waste utilization program. Waste waters from heavy summer runoffs in Tucson are being purified by passing them over grass runways before they are returned to pits or wells for replenishment of pumped ground water.

Perhaps one of the least appreciated pollution abatement processes is that of disease control through the soil. The soil accepts man's disease organisms in sewage, sludges, and leachates, and they disappear, or at least are inactivated. Most human and animal disease organisms cannot live in the highly competitive microbial activity of the soil. Management of cesspools and leaching fields permits the continuance of a healthy environment, even in crowded urban areas. However, with the increase in sewage from densely populated urban areas, the resulting sludges are becoming a greater disposal problem.

Air pollutants eventually reach the soil. Rain brings them down in a widespread and diluted state with little or no particular hazard to plant life. However, plants near the source of production of air pollutants can be adversely affected by their presence in the ambient air. The ozone (O_3) and peroxyacetyl nitrate in city smog, for example, are toxic to plants as well as man. Hydrogen sulfide and oxides of sulfur near mine smelters can become so concentrated that plant growth suffers. Millions of tons of objectionable air pollutants enter the soil each year, however, with no serious damage to plants or the soil.

One of the classic examples of environment improvement through soil management is that of the U.S. Department of Agriculture's and the State Agricultural Experiment Stations' long-standing joint program of soil conservation. Agricultural scientists have been aware of the need for an action program in soil and water control and management for many years. A conservation program has been ongoing since 1932. Agriculture has been and is active in many environment improvement endeavors.

Agricultural Experiment stations located in the desert Southwest began intensive research on the effect of radioactive elements from possible fallout of atomic bomb blasts as soon as this type of pollution occurred. Radioactive cesium and polonium control was studied at the University of California along with other uranium fission pollutants entering the soil, while control of strontium absorption by plants was studied at the University of Arizona.

Another historical program is that of organic matter utilization. Animal manures have been returned to the soil and used for fertilizers and soil conditioning since the beginning of man's organized growing of crops. Only since the concentration of animals in feeding pens, and with the change in agricultural chemical economy and the high cost of transportation, has disposal of manures been a problem. Like other wastes, dollar input is necessary. Manures do not pay back the cost of putting them on the land when much transportation is involved.

Complaints that commercial fertilizer contributes to contamination of underground water systems, streams, and lakes are proving to be less realistic than thought in the hysterical 1960s, in light of research data accumulation beginning about 1970. Phosphate, for example, is so tightly held to soil particles it does not move except as the soil moves. Nitrates are readily denitrified and lost to the atmosphere in many ways. Rarely can it be shown (except for shallow water tables) that fertilizer nitrogen moves into water recharge systems or accumulates in groundwaters in the desert Southwest.

Underground or well water sources evaluated for ammonia and nitrite nitrogen since 1889 by the University of Arizona show these waters contained nitrates 50 years before commercial fertilizers were used (see Table 10.1). Many of the early settlers who farmed the land and raised several generations of healthy families were unaware of the level of nitrates in their domestic and irrigation water.

Table 10.1

Nitrate Nitrogen Content of Some Groundwaters in Arizona*

Identification	Sampling date	Nitrogen (ppm) As nitrate (NO_3)	Nitrogen (ppm) As nitrogen (N)
	Published 1903**		
Alhambra (A. H. Smith)	9/16/1898	443	100
Cochise (S. Merill)	2/11/02	7	1.6
Prescott (City water supply)	3/10/00	25	5.6
Dragoon Mts. (Sorens)	9/ 5/01	206	46.3
Normal School	11/19/01	23	5.1
Phoenix (Thos. Murphy, Sec. 30)	4/22/02	120	27.0
Phoenix (2N, 4E)	4/22/02	103	23.3
Phoenix (Churchill)	9/15/1898	93	21.0
Phoenix (J. J. Harris, Sec. 6, 1N, 2E)	5/14/03	89	20.0
	Published 1964***		
T.2N, R.1W, Sec. 26	4/63	141	31.8
T.2N, R.1W, Sec. 27	2/60	454	102.5
T.2N, R.1W, Sec. 27	3/62	405	91.4
T.2N, R.1W, Sec. 28	4/63	425	95.9
T.2N, R.6W, Sec. 4	5/60	149	33.6
T.2N, R.6W, Sec. 13	1/52	217	49.0
T.2N, R.7W, Sec. 28	4/52	205	46.3
T.2N, R.8W, Sec. 31	9/60	9	1.8
T.2N, R.8W, Sec. 31	9/60	26	5.9
T.13S, R.24E, Sec. 35	9/63	25	5.6

*Data obtained from The University of Arizona Agricultural Experiment Station Soil and Water Testing Laboratory.

***From* W. W. Skinner. 1903. *The Underground Waters of Arizona — Their Character and Use*. Univ. Ariz. Agr. Expt. Sta. Bul. 46.

****From* H. V. Smith, G. E. Draper, and W. H. Fuller. 1964. *The Quality of Arizona Irrigation Waters*. Ariz. Agr. Expt. Sta. Report 223.

Fig. 10.5. Erosion of land resulting from mountain climbing activity of vehicles. Don't let it happen!

Nature herself has been an accumulator of nitrogen. Caves and geological deposits, as pointed out earlier for bat guano, contain natural sources of nitrate. Sources of native nitrogen appear more commonly in arid regions of the West and Southwest than in humid areas. High (native) nitrate soils are found extending from the deserts of Washington south into Mexico. Again these sources are unrelated to fertilizer use.

Much of the desert Southwest possesses that new, fresh look of virgin land unspoiled by man and so attractive to people who move into the area seeking clean air, fresh landscapes, and pleasant climate. Not the least attraction is the scenic beauty of rugged peaks cut into the sky, broad valleys, and exotically landscaped footslopes. Homes will be developed on these lands; towns and cities will stretch until they touch. Land use planning is essential to preserve the desert's beauty, yet allow adequate home development for the expected population increase.

As seen in Figure 10.5, this fragile beauty can readily be destroyed by nature enthusiasts ripping into the soil with their four-wheel-drive and other hill-climbing vehicles. As opposed to this, there is the conservation of beauty in the building of erosion control structures, dams, and lakes (Fig. 10.6). Birds, and indeed a wide variety of animal life, abound in such sanctuaries.

Fig. 10.6. Man-made pond adds beauty and a recreation facility to the landscape and helps control erosion as well.

This book was prepared to aid in preventing the mistakes of the past — in poor management of soils, water, and plantings — and to help build a pleasing landscape around homes and patios that will complement rather than insult the beauty nature provides so generously. This book can also remind us of our soil heritage, its fragile nature, its resilient capacity to recover under good management practices, and how and why soil management, so essential to the preservation of a high standard of living, can be compatible with the aesthetics of living in a nearly pollution-free environment.

Appendix Section

APPENDIX A
Organic Matter Cycles in Nature

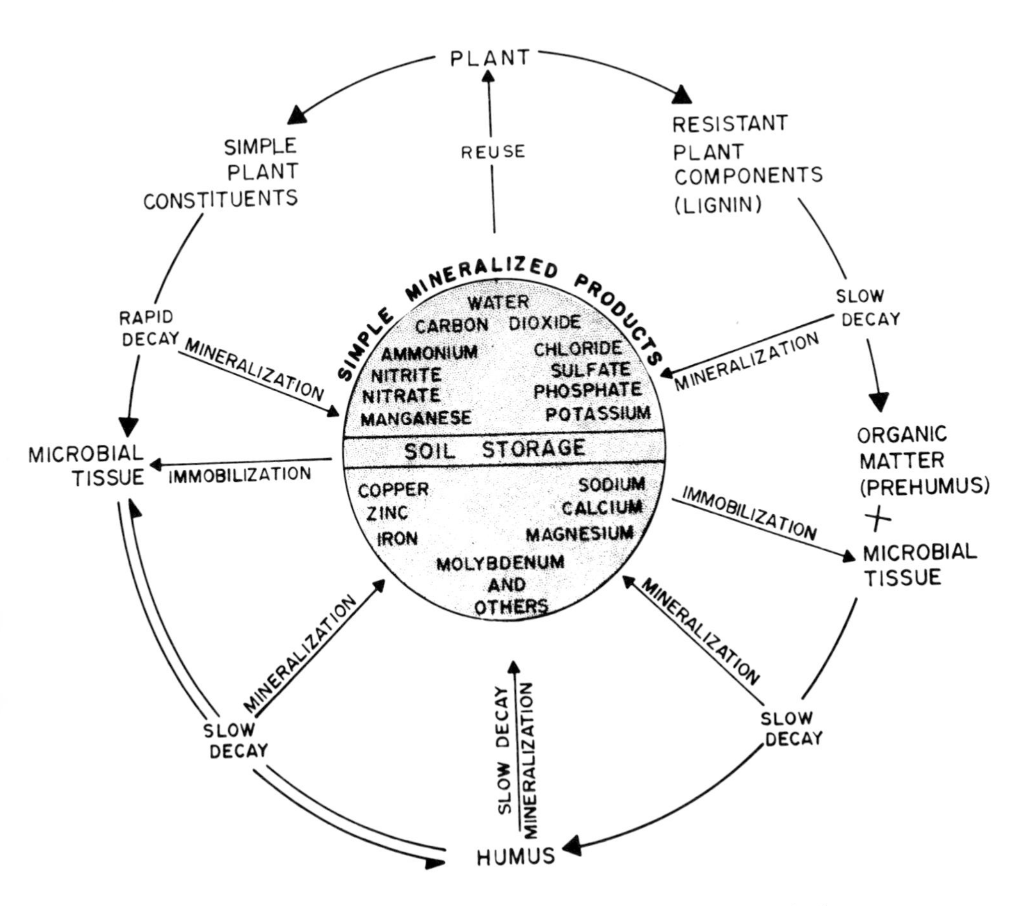

Fig. A.1. The organic matter cycle in nature showing decay, release of mineral nutrients, and humus formation.

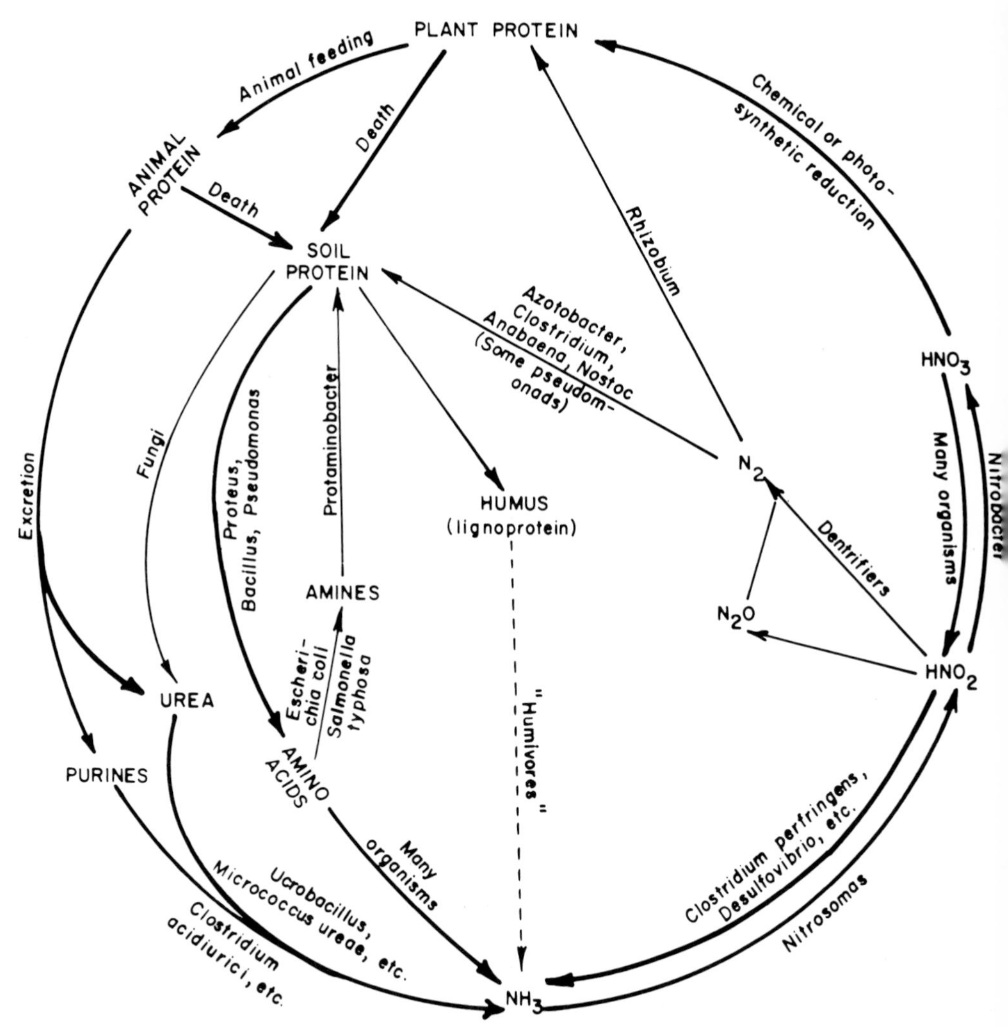

Fig. A.2. The nitrogen cycle in nature showing microbial transformations of nitrogen. (After K. U. Thimann. 1963. *The life of bacteria*. Ed. 2. Macmillan, New York.)

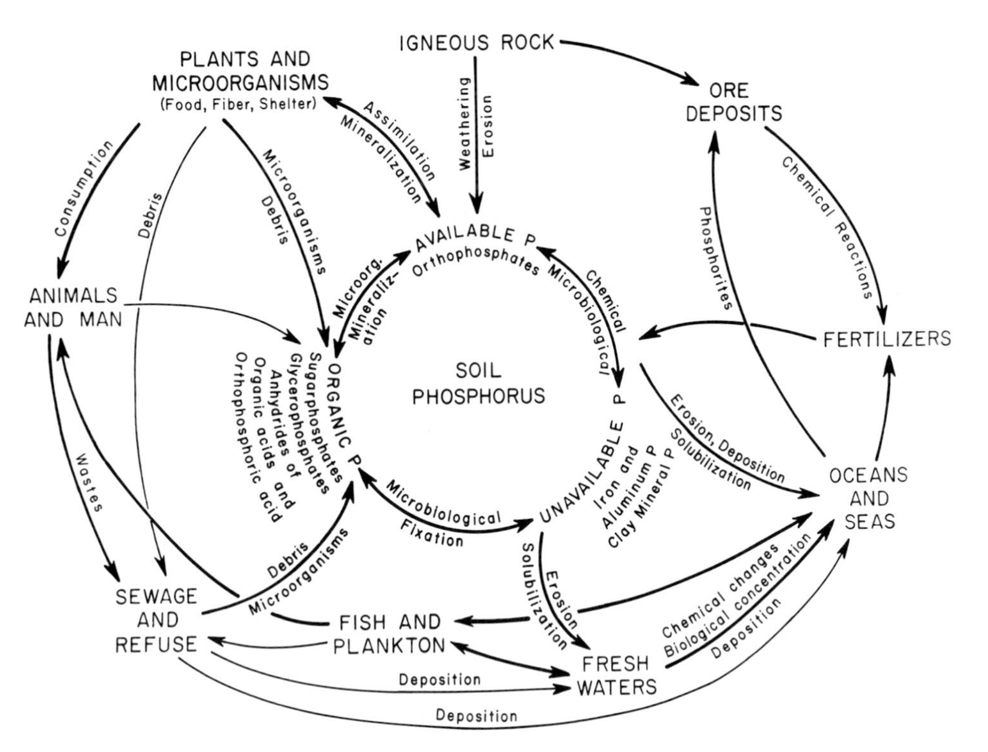

Fig. A.3. The phosphorus cycle showing its wide distribution. (After W. H. Fuller. 1973. "The phosphorus cycle," p. 951, *in* R. W. Fairbridge (ed.). *The encyclopedia of geochemistry and environmental sciences.* IVA. Van Nostrand and Rhinehold Co., New York.

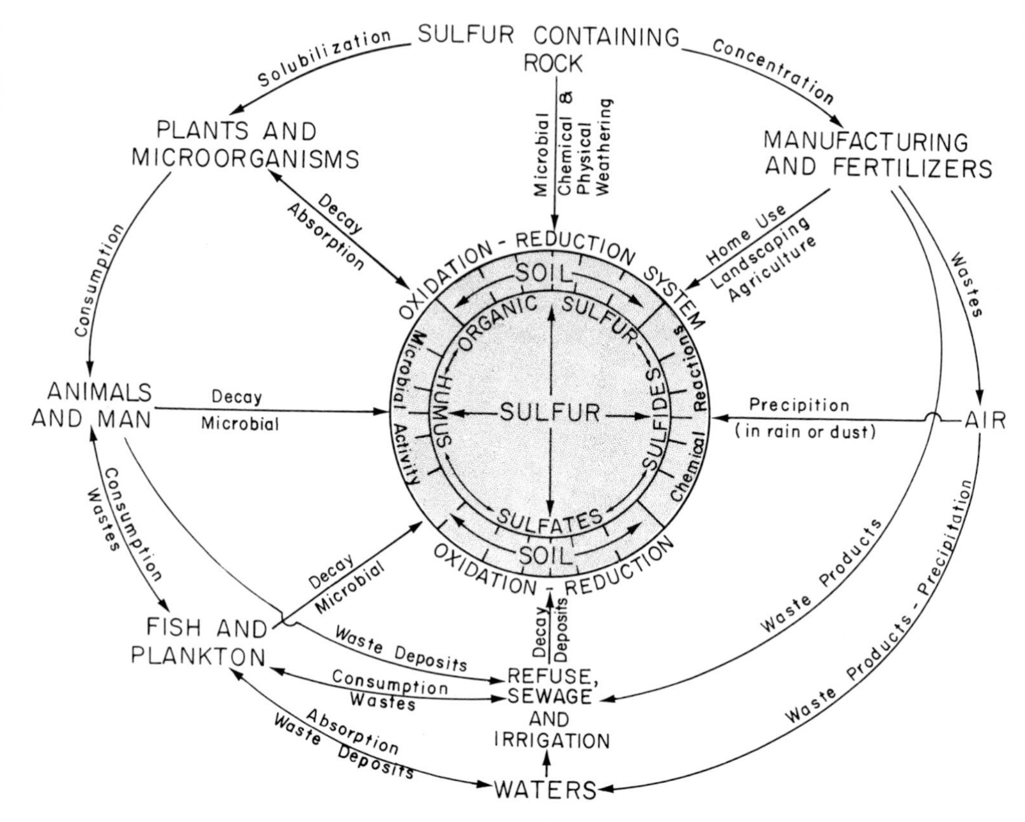

Fig. A.4. The sulfur cycle in nature showing oxidation of sulfur to sulfuric acid and sulfate formation and reduction to sulfides.

APPENDIX B

Conversion Factors for English and Metric Units

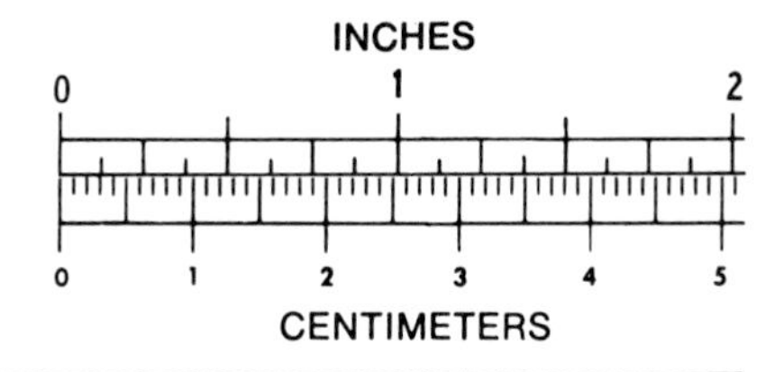

To convert column 1 into column 2, multiply by	Column 1	Column 2	To convert column 2 into column 1, multiply by
		Length	
0.621	kilometer, km	mile, mi	1.609
1.094	meter, m	yard, yd	0.914
0.394	centimeter, cm	inch, in	2.54
		Area	
0.386	kilometer^2, km^2	mile^2, mi^2	2.590
247.1	kilometer^2, km^2	acre, acre	0.00405
2.471	hectare, ha	acre, acre	0.405
		Volume	
0.00973	meter^3, m^3	acre-inch	102.8
3.532	hectoliter, hl	cubic foot, ft^3	0.2832
2.838	hectoliter, hl	bushel, bu	0.352
0.0284	liter	bushel, bu	35.24
1.057	liter	quart (liquid), qt	0.946
		Mass	
1.102	ton (metric)	ton (English)	0.9072
2.205	quintal, q	hundredweight, cwt (short)	0.454
2.205	kilogram, kg	pound, lb	0.454
0.035	gram, g	ounce (avdp), oz	28.35
		Pressure	
14.50	bar	lb/inch^2, psi	0.06895
0.9869	bar	atmosphere,* atm	1.013
0.9678	kg (weight)/cm^2	atmosphere,* atm	1.033
14.22	kg (weight)/cm^2	lb/inch^2, psi	0.07031
14.70	atmosphere,* atm	lb/inch^2, psi	0.06805
		Yield or Rate	
0.446	ton (metric)/hectare	ton (English)/acre	2.240
0.892	kg/ha	lb/acre	1.12
0.892	quintal/hectare	hundredweight/acre	1.12
1.15	hectoliter/ha, hl/ha	bu/acre	0.87
		Temperature	
	Celsius	*Fahrenheit*	
	−17.8C	0F	
$\left(\frac{9}{5}\,^\circ C\right) + 32$	0C	32F	$\frac{5}{9}(^\circ F - 32)$
	20C	68F	
	100C	212F	

*The size of an "atmosphere" may be specified in either metric or English units.

Index